T0295078

ROUTLEDGE LIBRARY EDITIONS: URBAN PLANNING

Volume 4

URBAN TRANSPORT PLANNING

URBAN TRANSPORT PLANNING

Theory and Practice

JOHN BLACK

Routledge
Taylor & Francis Group

LONDON AND NEW YORK

First published in 1981 by Croom Helm Ltd

This edition first published in 2018
by Routledge
2 Park Square, Milton Park, Abingdon, Oxon OX14 4RN

and by Routledge
711 Third Avenue, New York, NY 10017

Routledge is an imprint of the Taylor & Francis Group, an informa business

British Library Cataloguing in Publication Data
A catalogue record for this book is available from the British Library

ISBN: 978-1-138-49611-8 (Set)
ISBN: 978-1-351-02214-9 (Set) (ebk)
ISBN: 978-1-138-47839-8 (Volume 4) (hbk)
ISBN: 978-1-351-06860-4 (Volume 4) (ebk)

Publisher's Note
The publisher has gone to great lengths to ensure the quality of this reprint but
points out that some imperfections in the original copies may be apparent.

Disclaimer
The publisher has made every effort to trace copyright holders and would welcome
correspondence from those they have been unable to trace.

PREFACE

Urban Transport Planning: Theory and Practice – reissued

The re-issue of this book in 2018 coincides with 50 years of conducting transport research. In 1968, as an undergraduate at the Victoria University of Manchester, England, I collected field data on truck vehicle generation rates from industrial premises in Southeast London, calibrated a regression model and predicted what were, at best, primitive environmental and social impact assessments of freight movements. My doctoral research was on the topic of, what today we call, a "benchmarking study" of the modelling techniques of British land-use and transport studies undertaken in the 1960s and early 1970s. Its comprehensive international literature review (predominantly US research) covered best practice at that time.

When I gave my first university lecture to final year civil engineering students in January 1970 on the theory it was clear from student feedback that I needed to include more case studies, and this omission was again reinforced when I was appointed as a lecturer in 1975 at in the Postgraduate School of Transportation and Traffic at UNSW Sydney. Previously, I had been a Post-doctoral Fellow at the Australian National University where kindly mentors introduced me to the key players in government and consultancy in the field of land-use and transport planning. Some of these people are acknowledged in the original Preface. Whilst my graduate students appreciated the demonstration of how theoretical concepts (primarily from the USA) found their way into professional practice, it was Sir Alan Wilson on listening to a couple of my lectures at UNSW Sydney who suggested that a book should be structured into "theory" and "practice" (Canberra as a case study).

Over the years I have been approached variously to write a second edition to this book, to collaborate in an updated version in Japanese and Farsi (it was translated into Chinese in 1987) but my experience in writing second editions convinced me all of that was a lot of work and I should write a new book. I never did. Therefore, this offer from the publisher came as a pleasant surprise. I felt as a minimum contribution I should include here a few observations for teachers who may use the book as a basis for instruction, as well as for students, and where they should look for contemporary information.

Today, I continue to argue for the application of the relevance and power of the systems way of thinking when dealing with complex, and interrelated issues, such as urban transport and its impact on the economy, on society and on the environment. All quantitative models are based on assumptions and the contemporary relevance of this book (especially Chapter 3) are the explicit comments on the purpose of models as a *guide* to decision making, the structure of the model equations, and how data best supports the calibration of these models. The contemporary literature on modelling is now of wider international origin than when the research for this book was conducted, so instructors can readily design student exercises that incorporate recent theoretical developments. Also, the business case for any transport infrastructure project development requires the understanding of traffic forecasting methodologies because demand risk represents a significant part of all project risks.

Part Two of the book shows at varying spatial scales the application of such models in urban multi-modal transport planning practice. The Australian capital of Canberra is the case study but this is now nearly 40 years out of date. Readers are recommended to carefully study *Transport for Canberra Policy* (www.transport.act.gov.au), complemented by the ACT Planning Strategy, which will reinforce the contemporary importance of transport being linked to accessibility, social inclusion, health and well-being. Mees (2014) has summarised 100 years of transport planning in Canberra.[1]

The greatest changes in government approaches since the publication of this book are undoubtedly the inclusion of the environmental and social dimensions of transport, and greater stakeholder involvement in the formulation of project goals and objectives. In Australia, this is now recognised in national and state (territory) government legislation – referred to as "Ecologically Sustainable Development (ESD)." The other great innovation, not mentioned in the book, is the role of visualization, or, more specifically, in the transport context, geographical information systems (GIS). This has allowed a plethora of transport modelling tools as computer-based commercial software, from micro- to macro-level models to be developed over recent decades – all with their persuasive graphical outputs.

There are omissions in this book. Students might wish to pursue some of the following topics: sustainable transport planning; the development of airports and maritime ports; city logistics; micro-simulation models and agent-based modelling, including MATSim;[2] the financing of infrastructure, especially the modality of public–private partnerships (PPP); and technological improvements in vehicles and communications (intelligent transportation systems).[3] Despite all of the limitations of the book, it was written for students as an introduction to the fundamentals of land-use and transport interaction and how urban transport planning studies are conducted.

John Black
University of New South Wales, Sydney
9th November 2017

Notes

1 Mees, Paul (2014) A centenary review of transport planning in Canberra. *Progress in Planning*, 87, January, 1–32.
2 Horni, A., K. Nagel and K. W. Axhausen (eds) (2016) The Multi-Agent Transport Simulation MATSim. Ubiquity Press, London. DOI: http://dx.doi.org/10.5334/baw
3 www.ieeesplore.ieee.org

URBAN TRANSPORT PLANNING

THEORY AND PRACTICE

JOHN BLACK

CROOM HELM LONDON

© 1981 John Black
Croom Helm Ltd., 2–10 St John's Road, London SW11

British Library Cataloguing in Publication Data

Black, John
    Urban transport planning.
    1. Urban transportation — Planning
    I. Title
    711'.7'091732   HE305

    ISBN 0–85664–782–9
    ISBN 0–7099–0353–7 Pbk

Printed and bound in the United States of America

CONTENTS

TABLES

FIGURES

Figures

TO MY PARENTS

PREFACE

Since 1970, I have given lecture courses in transport planning, at under-graduate and post-graduate levels, to engineers and planners at Bradford University, England, the University of New South Wales, Australia, and the Asian Institute of Technology, Thailand. A common reaction of students when confronted with methodology—especially those parts involving quantitative methods—is to ask about its practical application and relevance. Originally, my intention was to write a case study of Canberra's transport planning, but Professor Alan Wilson encouraged me to combine this idea with material from some of my lecture notes to produce a book on both the theory and the practice of urban transport planning.

Any thorough investigation of the physical development of cities, of which transport planning is one agent of change, crosses traditional disciplinary boundaries such as engineering, town planning, economics and geography. With this in mind, I have tried to draw together the pertinent material into a convenient form that makes it suitable as an introductory textbook on transport planning methods. It is written primarily for students of civil engineering, town planning and urban geography, but others who are concerned with the impact of transport either on society or on the environment should find parts of the book useful. For those who may want to pursue the subject in more depth, the references made in the text to a substantial body of technical literature aim to provide material for further study.

In writing this book, I should like to acknowledge four main influences. First, Mr Peter Harrison, former Director of Town Planning, National Capital Development Commission, Canberra, has been a continuing source of help and encouragement. Second, Professor Ross Blunden, Head of the School of Transport and Highways at the University of New South Wales, initially as a teacher, and later as a colleague, has given me theoretical insights into the subject. Third, Professor Bruce Hutchinson, Department of Civil Engineering, University of Waterloo, provided me with the basic idea for structuring different types of transport planning activity by their geographical scale and their time horizon. Finally, my students have contributed by asking critical questions and by commenting on early outlines of chapters.

The task of revising a first draft was made easier by the generous

Preface

help from my former colleagues, Mr Peter Harrison and Dr Ian Manning, at the Urban Research Unit, Australian National University, and from Professor Alan Wilson, University of Leeds. Dr Coleman O'Flaherty, former First Assistant Commissioner for Engineering with the National Capital Development Commission, has provided me with valuable material. The chapter on public transport has benefited by comments made by Professor John Shortreed, University of Waterloo, and the section on area traffic control was greatly improved by suggestions made by Dr Rahmi Akcelik, of the National Capital Development Commission.

I am especially indebted to my wife, Ms Charmian Gaud, for typing the manuscript and for making valuable suggestions on how to improve the text. Most of the illustrations were drawn by Ms Gail Saunders; Mr Keith Mitchell drew two maps of Canberra. My appreciation extends to all of these people, but any errors remain my responsibility.

John Black,
School of Transport and Highways,
University of New South Wales.

INTRODUCTION

This is a book about the methodology of transport planning, and it touches only briefly on transport policy. It is written in the belief that an understanding of land use, traffic and transport interaction is necessary if transport plans and policy decisions are to be effective and achieve their aims. Whilst recognising that planning is ultimately a political matter, the value of urban transport planning methodology is that in an ideal sense it represents a way of gathering and presenting information to decision-makers detailing who is affected by land-use changes and transport improvements and to what extent.

The methods or procedures of urban transport planning can be organised conveniently using the systems approach, a conceptual tool widely used in the study of physical and social systems which enables complex and dynamic situations to be understood in broad outline. Here, the system of interest is urban land use and transport, and it is the interactions within this system that give rise to the phenomena of traffic. The emphasis is on quantitative analysis: it is argued that first an understanding of the relationships between land use, traffic and transport must be established before attempting to devise solutions.

The structure of the book is as follows: it is divided into theory and practice. Part One explains the application of systems modelling in transport analysis and forecasting which forms the theory of urban transport planning. The contents can be read given a knowledge of elementary mathematics, including, in particular, algebraic functions which utilise subscripts and superscripts in their notation. Part Two is a more descriptive presentation of urban transport planning practice. It recognises that planning involves *time* horizons which range from short- to long-term, and *geographical* scales which range from metropolitan-wide down to local areas, and that the systems approach is suitably flexible to such a wide range of applications.

The book opens with an explanation of the systems approach and an introduction to the fundamentals of land-use—transport interaction. Definitions of land use, traffic and transport are given and the way that this system functions in theory is described in both words and with mathematical notation. The next three chapters build upon these fundamentals.

Chapter 2 outlines transport data collection methods and uses this information to describe urban transport modes in terms of their operational performance and carrying capacity. For analytical convenience, transport is represented as a network of nodes and links; its performance is measured by travel times or costs along each link.

Chapter 3, on travel demand analysis, is the longest, covering data collection methods and aggregate and disaggregate models. Aggregate travel demand is divided into zonal traffic generation, traffic distribution, modal split and traffic assignment, whereas disaggregate analysis deals with individual travel behaviour. Each model is taken in turn and its theoretical structure is given, appropriate calibration procedures are explained and validation methods are outlined.

Chapter 4 considers forecasting methods, plan preparation and evaluation procedures. Because traffic is a function of land use, forward projections of population, employment and car ownership are required before future travel demand can be established. The steps and assumptions in making traffic forecasts are indicated and the likely accuracy of some of the forecasts are discussed. Principles used in the formulation and testing of alternative transport plans are indicated and methods for their economic, social and environmental appraisal are introduced.

Part Two demonstrates practical applications of the systems approach and of the theory. Chapter 5 describes the conventional transport study, which has formed the backbone of long-term urban planning from the mid-1950s onwards. Strategic transport planning is considered in Chapter 6, which explains the procedures for calculating the traffic implications of radically different urban development options. Chapter 7, on public transport, concentrates on passenger forecasting methods, the appraisal of technologies and the use of attitudinal surveys to indicate how to improve existing services.

The next two chapters describe planning approaches which have a much shorter time horizon. Chapter 8 describes urban traffic management, such as area traffic control and bus priority schemes. It also illustrates how behavioural travel demand models may be applied in examining the implications of different urban transport policies. Chapter 9 concentrates on some very location-specific transport planning problems, including the provision of pedestrian and parking facilities, the traffic effects of isolated land-use developments, and streets and cycle paths in residential areas. Finally, Chapter 10 gives a general overview of the strengths and weaknesses of transport planning practice.

Within Part Two, examples of these various types of transport planning study are drawn from Canberra, the national capital of Australia. There are a number of reasons why Canberra has been selected as a case study. First, the National Capital Development Commission (NCDC) has the statutory responsibility to plan and develop Canberra, and a fair measure of co-ordination is achieved between land use and transport planning. Second, Canberra, like many cities, has experienced the pressures of rapid urban growth (with the population increasing from about 35,000 in the mid-1950s to almost 250,000 by 1980) which has necessitated the advance planning of suitable transport facilities. Third, the NCDC has a policy of commissioning professional advice from consulting firms, whose reports provide well documented evidence on planning procedures, which is not always the case in other cities. Finally, Canberra is a new town and it is easier to see the influence of planning studies on the resultant pattern of urban development.

PART ONE: THEORY

It is a sin to think that
calculation is invention . . .

    Goethe (cited in Leibbrand, 1970, p. 8)

1 FUNDAMENTALS OF LAND-USE−TRANSPORT PLANNING

Many of the problems facing urban societies are complex and not altogether obvious: traffic and transport problems are clearly prominent, confronting the public on a day-to-day basis. There is a wide divergence of opinion on how to solve the 'urban transport problem', but the aim of transport planning is to search for the best solutions given the resources available.

Transport planning, as a professional activity, can be justified to the community only if problems and solutions are considered in a rigorous way, including a detailed analysis of all relevant factors. This chapter provides an introduction to the fundamentals of transport planning and argues strongly for the need to understand how a city works in terms of the interaction between land use, traffic and transport.

The systems approach provides the planners with a suitable framework for pursuing these ideals. In the first section, land use, traffic and transport are defined as a 'system' and the steps of the transport planning process are outlined. The next two sections explain how the system 'works', first following a descriptive approach, then following a quantitative approach. Systems modelling is best explained with simple worked examples and these are presented in section four. The final section indicates the application of systems modelling in the preparation of alternative plans.

1.1 A Systems Planning Framework

The major steps that make up an orderly approach to planning are stated in Figure 1.1. This framework is perfectly general, and has found numerous applications in planning (McLoughlin, 1969; Chadwick, 1978). A clear statement of the problem is necessary before the purpose or objectives of a study can be specified. Steps 2 and 3 suggest that an understanding of how any system works is based on reliable data and analytical methods. Quantitative methods are used in step 4 to forecast how the system might evolve in the future. The uncertainty surrounding the future is recognised in step 5 by examining alternative plans. Step 6 specifies the criteria and procedures for choosing the best plan.

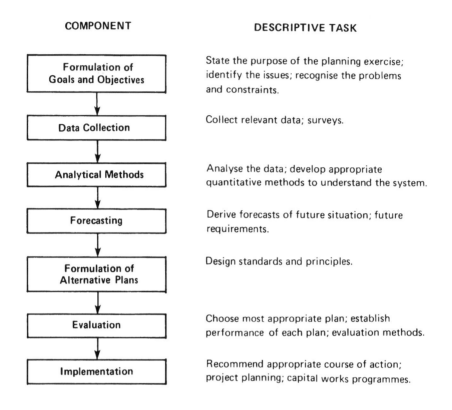

COMPONENT   DESCRIPTIVE TASK

Formulation of
Goals and Objectives   State the purpose of the planning exercise;
                            identify the issues; recognise the problems
                            and constraints.

Data Collection   Collect relevant data; surveys.

Analytical Methods   Analyse the data; develop appropriate
                            quantitative methods to understand the system.

Forecasting   Derive forecasts of future situation; future
                            requirements.

Formulation of   Design standards and principles.
Alternative Plans

Evaluation   Choose most appropriate plan; establish
                            performance of each plan; evaluation methods.

Implementation   Recommend appropriate course of action;
                            project planning; capital works programmes.

Figure 1.1: The Systems Approach

Finally, plan implementation requires resources and political support. This linear progression is a little misleading, because the systems approach contains feedback loops from one step to another to ensure internal consistency.

The word 'system' is often used loosely, but it does have a precise scientific meaning. A system is 'a set of objects together with relationships between the objects' (Hall and Fagen, 1956, p. 18). This is based on the observation that in any organised system of component parts (objects) the behaviour of any one part has some effect on or interaction with other parts. In transport analysis, the system is comprised of three major objects or components:

(a) *Land Use*–deceptively simple words that convey a complexity of meanings. In this book, they are used very broadly to mean (i) the legal use to which the land is put (residential, industrial, etc.);

(ii) the type of structures built on the land (houses, factories, schools); and (iii) measures of the intensity of social and economic activities that take place on the land (population, employment, factory output, etc.).
(b) *Transport Supply*–forms the physical channels or links between land use. It includes (i) a variety of transport modes such as footpaths, roads, tramways, bus routes and railways; and (ii) the operational characteristics of these modes, such as travel times, costs or service frequencies.
(c) *Traffic*–is the joint consequence of land use *and* transport supply. Pedestrian and vehicular traffic represents the horizontal movement of people and goods over the transport network.

1.2 Understanding the System–a Description

Understanding how the system works is closely tied to the unravelling of the interactions between land use, traffic and transport supply. Five concepts are fundamental:

(a) accessibility;
(b) traffic generation;
(c) spatial pattern of traffic;
(d) selection of transport mode and route; and
(e) traffic on the transport network.

1.2.1 Accessibility

Accessibility is the concept which combines the geographical arrangement of land use and the transport that serves these land uses. Accessibility is a description of how conveniently land uses are located in relation to each other, and how easy or difficult it is to reach them via the transport network. Figure 1.2 presents a simple scheme for classifying accessibility. When many land-use activities are located close together and the transport connections are good, high accessibility is achieved. Conversely, when activities are located far apart and the transport connections are poor, low accessibility results.

Different geographical locations do not have the same accessibility because land-use activities are distributed unevenly and transport is neither of uniform coverage nor quality. Some land uses have a dispersed pattern (e.g. dwellings), others are more clustered (e.g. shops), and a

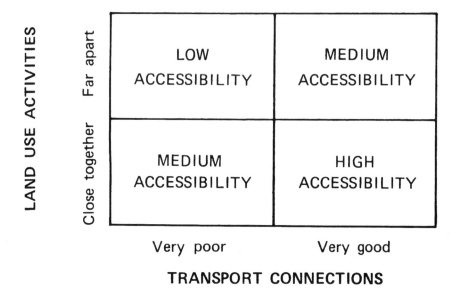

Figure 1.2: Classification of Accessibility Levels

few special activities have 'one-off' locations (e.g. airports). Inner-city public transport is usually better, especially when connecting the city centre.

Accessibility also provides a measure of the performance of the land-use–transport system. Residents are primarily interested in accessibility to job opportunities, schools, shops, health services, leisure and recreation activities. Retailers are concerned about accessibility to customers. Industrialists rely on accessibility to labour markets and to suppliers of materials.

1.2.2 Traffic Generation

Traffic is a function of land use. The potential of a block of land to generate traffic is realised when activities take place on the land. Traffic generation of a block of land is a measure of the amount of traffic (number of people or vehicles, tonnes of freight) that visits during a specified time period (usually per day, or per hour). An example is the number of people going into and coming out of an office building each day.

The amount of traffic generated is related to: (a) the type of land use; and (b) the scale or intensity of the activity taking place in the land. Traffic generated by each land-use category is a reflection of its role in the social and economic functioning of the economy. For example, the average weekday number of vehicular trips generated by one hectare of commercial land is ten times higher than for one hectare of residential land with single-family detached houses; fast food restaurants generate ten times the number of vehicle trips per thousand gross square feet than quality restaurants (ITTE, 1976). As the scale or intensity of land-use activity increases, so too does the amount of traffic generated.

1.2.3 Spatial Pattern of Traffic

The spatial pattern or distribution of traffic is explained by two factors: (a) the disposition of land use; and (b) the restraints on movement— distance acts as a friction to movement. Interaction between two complementary but spatially separated land uses involves a movement of people, goods or information. The spatial separation represents a 'barrier' to interaction, and there is a general preference for short-distance movements rather than long-distance movements. The amount of traffic measured between any two places depends on the land-use intensity in both places and the frictional effect of distance. Figure 1.3 illustrates that the amount of traffic interaction is related positively to

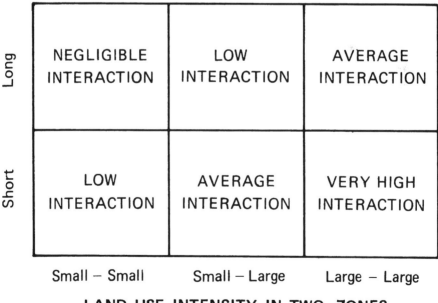

Figure 1.3: Classification of Traffic Interaction Levels

the intensity of land use and negatively (or inversely) to increasing distance between two places. Interaction is very high when two land uses with intensive development are close together. Conversely, interaction is negligible when two land uses with low activity levels are far apart. These are extreme situations and intermediate levels of traffic interaction are of course possible, and more likely.

1.2.4 Selection of Transport Mode and Route

When interaction between two land uses does take place, travellers, or shippers responsible for moving freight, must select which transport mode and what particular route to follow. The journey itself has unattractive aspects, including cost, time, discomfort, danger and uncertainty, and those modes or routes which are least unattractive will be preferred. A traveller will patronise the transport mode (or combination of modes) and route which takes the shortest travel time or costs the least from origin to destination. The choice is based on the desire to minimise the friction in overcoming distance. This assumes that the traveller or shipper has perfect information about the alternative modes and routes and then acts rationally in selecting the best way of getting about. It also assumes that people have a genuine choice: if not, they are said to be 'captive' to one transport mode or route.

1.2.5 Traffic on the Transport Network

As traffic increases on a mode or route the difficulty in getting about also increases, especially when traffic approaches the fixed capacity of the transport facility. Crowded buses mean waiting in queues; congested roads mean stop-start conditions for motorists; and crowded footpaths force pedestrians to shuffle along at a pace slower than normal walking speed.

Travel times are an indicative measure of the difficulty in getting around on transport. The travel times on transport networks vary with traffic flow. This relationship is non-linear: as traffic increases from zero travel times increase only slightly, but as traffic approaches capacity there are large increases in travel time with each increment of traffic.

1.3 Understanding the System–Systems Modelling

Most disciplines which are concerned with deeper levels of explanation of how things work progress from a descriptive to a quantitative approach, which in systems terms is called 'systems modelling'. The relationships between land use, traffic and transport described above can be expressed also as systems models.

1.3.1 Model Definition

A 'model' is a representation of something else and is designed for a specific purpose. When the purpose is to remind ourselves what that 'something else' looks like, the representation might take the form of a scale model, a photograph or a map. When the purpose is explanation of something that is quite complex, the model need not necessarily 'look like' whatever it represents; instead it may take the form of mathematical equations. All models share one common characteristic irrespective of their intended purpose: the mapping or transformation of the real world into a 'model'. When every feature is retained the model becomes a replica but usually abstractions, simplifications or generalisations are made from the real world. The mathematical model is highly abstract because our perception of the real world is simplified and translated into the 'language' of mathematics.

> Mathematical notation is a more precise language than English. Because it is less ambiguous, a mathematical model is a description which has greater clarity than most verbal models . . . There is nothing inherent in the symbols of mathematics which guarantees accuracy. However, the precision required to translate words into symbols can often reveal inadequacies in the verbal description, and may thus lead to a sharpening-up or clarification of our mental image of the way in which we think the real-world system operates (Lee, 1973, p. 8).

1.3.2 Glossary—Systems Modelling

The following glossary may be helpful because technical terms commonly used in systems modelling occur in this book.

Notation: the most elementary algebra consists of forming expressions with letters, which stand for numbers or a range of numbers, and with the basic operations of ordinary arithmetic—addition, subtraction, multiplication and division. Standard symbols represent arithmetic operations but the choice of notation is somewhat arbitrary. Consistency is important so in this book, *variables* are indicated by upper-case Roman letters, and *parameters* or *coefficients* by lower-case Greek letters.

Function: a mathematical concept used whenever the value of some quantity (dependent *variable*) is regarded as dependent on, or determined by, another quantity or quantities (independent *variables*).

Argument: denotes what a *function* is a function of, i.e. the independent variables. Particular values of the function can be calculated by assigning values to the variables which occur in its expression, or argument.

Variable: a quantity that is able to assume different numerical values. If a letter is used to denote the value of the *function* it is called the dependent variable. If letters are used for the *argument* of the *function*, they are called independent variables.

Parameter: a quantity which is constant in each case under consideration, but which may vary from case to case. The *functions* for the different cases may all have the same general *argument* of *variables* and parameters, but the value of the parameters is unique to each particular case.

Coefficient: in mathematics and its applications, a term synonymous with *parameter.*

Calibration: the process of estimating the *parameters* or *coefficients* to get the best correspondence between the model predictions and data representing the real world.

Algorithm: a procedure for performing a complicated arithmetic operation by carrying out a precisely determined sequence of simpler ones. These procedures form suitable subjects for computer programmes.

Those readers in need of a short refresher in mathematics are referred to Lee (1973, pp. 28–40) or to Wilson and Kirkby (1975, pp. 3–103), where the necessary mathematical preliminaries essential to an understanding of urban models are explained. A check-list of factors to consider when designing any type of model is given by Wilson (1974, p. 31).

In transport analysis, the *purpose* behind a systems model that represents land use, traffic and transport is twofold: to gain a better understanding of how the system works; and to predict the traffic consequences of planned changes to land use and transport.

1.3.3 Land-use–Transport Systems Model

The land-use–transport systems model contains three quantifiable variables: land use (number of people, number of jobs, income, car ownership characteristics of the population), traffic (number of passengers, number of vehicles) and transport (travel times, costs). Traffic is the dependent variable except in the calculation of transport travel times when it becomes an independent variable. Land use is a variable because its intensity varies in different parts of an urban area and because it changes over historical time. Transport is a variable because its quality and quantity vary geographically and because it too changes over time – the building of new roads, the extension or contraction of public transport services. Each variable in the model is identified by the notation: L = land use; T = transport supply; Q = traffic.

As the ultimate aim is to represent the amount of traffic in the urban area, the variables require identification in a location-specific way. The urban area is partitioned into a set of discrete geographical zones with the amount of zonal land-use activity assumed to be located in one place called the centroid. The number of zones chosen by the analyst is a question of aggregation, but obviously greater precision is obtained with a large number of small zones. The major transport facilities are represented by networks connecting the zone centroids.

Representing land use and transport in this way has analytical advantages because the system can be specified in numerical terms.

(a) A map reference (grid co-ordinate) defines the exact position of each zone centroid.
(b) A numerical code is used instead of a place-name to identify the geographical location of a zone.
(c) Different measures of zonal land use can be tabulated and cross-referenced with the zone numbering scheme.
(d) The operational characteristics of transport – travel times or costs – are attached to each link of the transport network. Inter-zonal travel time is obtained by adding times on the sequence of links from an origin zone centroid to a destination zone centroid.
(e) The amount of traffic produced by or attracted to each zone centroid is measured. Traffic flow over the network is also measured.
(f) A numerical code is used to identify the position of the transport facilities. Separate coding schemes may be developed for each transport mode.

The conventional notation is to refer to any origin zone as zone i and any destination zone as zone j. Variables associated with any zone are labelled with the subscripts i or j, or both. A journey has an origin *and* a destination, so two subscripts are necessary for some variables:

L_{oi} = land use in the origin zone i;
L_{dj} = land use in the destination zone j;
T_{ij} = transport travel time or cost from zone i to zone j;
Q_{pi} = total traffic produced by zone i;

Q_{aj} = total traffic attracted to zone j;
Q_{ij} = traffic from zone i to zone j;
Q_k = traffic flow on route k.

This notation is used to indicate how the descriptive interactions between land use, traffic and transport are translated into a conceptual model of the system. Additional notation is introduced when necessary.

The *accessibility* of an origin zone is an indication of how convenient land-use activities are located in relation to that zone, and how easy or difficult it is to reach them via the transport network. The total accessibility of zone i to any nominated activity in all destination zones j, including those activities located in the origin zone, is a function (f) of land-use intensity and transport supply:

$$H_i = f(L_{dj}, T_{ij}) \qquad (1.1)$$

where

H_i = accessibility of zone i to a nominated land-use activity.

Traffic generation is a function of land use. The amount of traffic produced by an origin zone i is proportional to the type and to the intensity of land use in zone i:

$$Q_{pi} = f(L_{oi}) \qquad (1.2)$$

Some recent hypotheses on traffic generation include accessibility as an independent variable. The amount of traffic attracted by a destination zone j is proportional to the type and intensity of land use in zone j:

$$Q_{aj} = f(L_{dj}) \qquad (1.3)$$

Different measures of land use are appropriate for all origin zones and for all destination zones. Zonal traffic generation is the sum of traffic production and traffic attraction.

The *spatial pattern of traffic* amongst zones is a function of the land-use intensity in zone i, which produces traffic, the land-use intensity in zone j, which attracts traffic (different measures of land use are appropriate for all origin zones and for all destination zones), and transport difficulty of getting from one zone to another:

$$Q_{ij} = f(L_{oi}, L_{dj}, T_{ij}) \qquad (1.4)$$

The *selection of transport mode and route*, for those origin-destination journeys where there is an element of choice, is based on a comparison of the operational characteristics of the competing transport modes and routes. Mode selection, in the simplest case of a choice between public transport ($m=1$) and private transport ($m=2$), is:

$$Q_{ij(m)} = f(T_{ij(1)}, T_{ij(2)}) \qquad (1.5)$$

where the traffic carried on the alternative modes is identified by the additional subscript m. Similarly, mode-specific route selection–the sequence of network links followed from an origin zone centroid– is based on a comparison of the operational characteristics of alternative transport routes for each mode:

$$Q_{ij(k)} = f(T_{ij(1)}, T_{ij(2)}, \ldots T_{ij(k)}) \qquad (1.6)$$

where the traffic carried on each alternative route is identified by the subscript k. Here, the implicit (m)-label has been dropped for simplicity. Thus, $T_{ij(1)}$ in this equation should be carefully distinguished from $T_{ij(1)}$ in equation (1.5). If necessary, the notation $T_{ij(m)(k)}$, for the travel time of the kth route in mode m, could be used.

The amount of *traffic on each transport route* determines the transport travel times (and user costs), especially on roads. Thus,

$$T_k = f(Q_k) \qquad (1.7)$$

where the subscript k identifies the transport network route followed by the traveller.

The non-linear relationship between traffic flow-dependent travel times on transport facilities has been studied extensively, and one of the possible algebraic functions with the required theoretical properties (Blunden, 1971, pp. 80-4) is:

$$y = (1 - \alpha x)/(1 - x) \qquad (1.8)$$

This is plotted graphically for three selected values of the parameter α (Figure 1.4). The y-axis can be interpreted as the ratio of the travel time for a range of traffic loads to the minimum travel time with no traffic load; and the x-axis as the ratio of traffic flow to saturation flow (transport capacity), which is a measure of the traffic load. The gradient of the function is determined by the parameter.

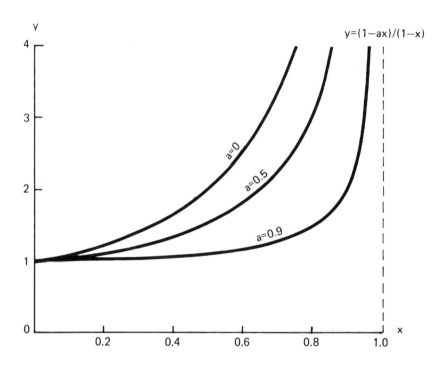

Figure 1.4: Properties of a Traffic Flow–Transport Impedance Function

Conceptually, the equations of the system are the travel demand and transport supply. Equations (1.2) to (1.6) are the amount of travel by location, by geographical pattern and by transport mode and route, whereas equation (1.7) is the mode-specific supply function. The solution to any set of equations with specified parameters is not straightforward because both traffic and transport supply variables are found on the left- and on the right-hand sides of the model. Continuous adjustments in traffic and transport supply are required to reach a solution that balances in all equations. Such a solution is called the *equilibrium solution* to the problem.

Under heavy traffic conditions there is a spread, or diversion of vehicles, over alternative transport routes, such that travel times become equal:

$$T_1 = T_2 = \ldots = T_k \tag{1.9}$$

This is based on the traffic flow principle which holds that no individual traveller can become better off by finding a route with travel times lower than those defined by an equilibrium, or an equal travel time, assignment (Wardrop, 1952). Equilibrium is an important concept, which is best clarified by studying the following simple example.

1.4 Land-use–Transport Interaction–an Example

Two zone centroids are connected by a main arterial road (route 1) and an alternative route along local streets (route 2). The focus for illustrative purposes is on one trip interchange only, between zones 1 and 2. Usually, of course, there will be many of these. Zone 1 is a residential area with 30,000 people and zone 2 is an employment centre with 10,000 jobs. In the following equations the measure of land-use intensity is $L_{o1} = 30,000$ and $L_{d2} = 10,000$. The equations and parameters below form the systems model of land-use–transport interaction. It is assumed that all travellers use one mode.

Accessibility

$$H_{12} = L_{d2}/T_{12} \qquad (1.10)$$

where

H_{12} = the accessibility of zone 1 to the employment opportunities located in zone 2 in jobs reached per minute;

T_{12} = travel time in minutes from zone 1 to zone 2.

(More generally, later, it will be shown that T_{12} may be raised to some power.)

Traffic Generation

$$Q_{p1} = 0.4\,L_{o1} \qquad (1.11)$$

$$Q_{a2} = 1.0\,L_{d2} \qquad (1.12)$$

where

Q_{p1} = peak-hour number of vehicle trips produced by zone 1;

Q_{a2} = peak-hour number of vehicle trips attracted to zone 2;

L_{o1} = land-use activity of zone 1 (i.e. population);

L_{d2} = land-use activity of zone 2 (i.e. employment).

Spatial Pattern of Traffic

$$Q_{12} = 0.001\,Q_{p1}.Q_{a2}/T_{12} \qquad (1.13)$$

where

Q_{12} = peak-hour number of vehicle trips from zone 1 to zone 2.

Flow-dependent Travel Times

$$T_{k(Q)} = T_{k(0)} \cdot \frac{1 - (1 - \lambda)\, Q_k/Q_{max(k)}}{1 - Q_k/Q_{max(k)}} \qquad (1.14)$$

where

$T_{k(Q)}$	=	the travel time in minutes on route k at vehicular flow Q;
$T_{k(0)}$	=	the travel time in minutes on route k at 'zero' traffic flow;
λ	=	level of service parameter associated with each route;
Q_k	=	traffic flow (vehicles per hour) on route k; and
$Q_{max(k)}$	=	saturation flow (vehicles per hour), or transport capacity, of route k.

Table 1.1 gives the transport supply characteristics for the two roads.

Table 1.1: Transport Supply Characteristics for a Simple Network

Transport Supply Characteristics	Route	
	$k = 1$	$k = 2$
Length in kilometres	16	19
'Zero-flow' travel time in minutes, T_0	24	38
Level of service parameter, λ	0.3	1.0
Saturation flow, Q_{max} in veh/hour	3,000	2,000

1.4.1 The Problem and Solution

The problem is to find the equilibrium solution to the system of equations. This involves calculating: (a) the total peak-hour traffic flow from zone 1 to zone 2; (b) the amount of traffic using each route, assuming that traffic 'satisfies' Wardrop's principle of an equal travel time assignment; and (c) the inter-zonal travel time which will be identical on either route. The problem may be solved by a graphical analysis or by algebra.

The steps in the graphical approach are as follows. From equation (1.11) the zonal amount of traffic produced by zone 1 is:

$$Q_{p1} = 0.4 \times 30,000 = 12,000 \text{ vehicles per peak hour.}$$

From equation (1.12) the zonal amount of traffic attracted to zone 2 is:

$$Q_{a2} = 1 \times 10,000 = 10,000 \text{ vehicles per peak hour.}$$

Substituting these into equation (1.13) the inter-zonal pattern of traffic is:

$$Q_{12} = 120{,}000/T_{12} \text{ vehicles per peak hour.} \qquad (1.15)$$

The function for the traffic flow-dependent travel times on route 1 is obtained by substituting the appropriate values from Table 1.1 into equation (1.14). Simplifying the expression:

$$T_{1(Q)} = 24\,(3{,}000 - 0.7\,Q_1)/(3{,}000 - Q_1) \text{ minutes.} \qquad (1.16)$$

The traffic flow-dependent travel times for route 2 are found in an identical way, and simplification gives:

$$T_{2(Q)} = 76{,}000/(2{,}000 - Q_2) \text{ minutes.} \qquad (1.17)$$

Equations (1.15), (1.16) and (1.17) are solved for a range of values of travel time and traffic flow substituted into the right-hand side and plotted graphically in Figure 1.5. The vertical axis of the graph is travel time in minutes and the horizontal axis is traffic flow in vehicles per hour. The upward-sloping flow-dependent curves plotted separately for the two routes indicate that travel times increase with additional traffic, whereas the downward-sloping curve (dashed line) for inter-zonal traffic demand decreases with an increase in travel time. The traffic flow-dependent travel times for the transport corridor are the sum of the curves for route 1 and route 2, as shown by the dotted and dashed line.

The point where the traffic flow-dependent travel time curve for the *transport corridor* intersects the inter-zonal traffic demand curve represents the equilibrium solution. From the graph the value obtained is a total flow of 2,610 vehicles per hour at a travel time of 46 minutes. A line drawn from the equilibrium point parallel with the horizontal axis intersects the curves for route 1 and for route 2 to give a traffic flow of 2,260 vehicles per hour on route 1 and 350 vehicles per hour on route 2.

The steps leading to an algebraic solution follow. In order to simplify the equations traffic flows and saturation flows (transport capacity) are expressed in units of 1,000. Thus, equation (1.15) now becomes:

$$Q_{12} = 120/T_{12} \text{ vehicles per peak hour in } 1{,}000\text{s} \qquad (1.18)$$

equation (1.16) now becomes:

Figure 1.5: Graphical Solution to Land-use—Transport Interaction Problem—the Existing Situation

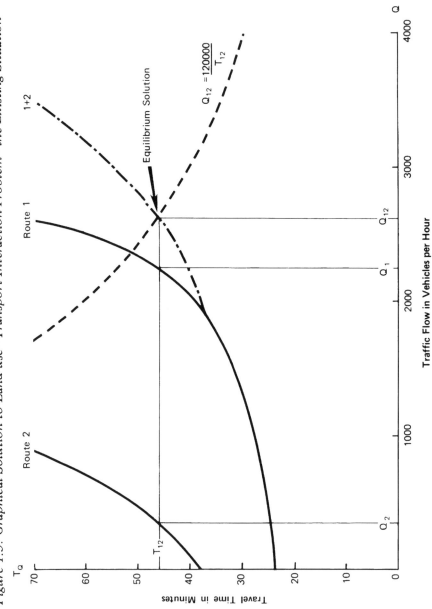

$$T_{1(Q)} = 24\,(3 - 0.7\,Q_1)/(3 - Q_1)\ \text{minutes} \qquad (1.19)$$

and equation (1.17) now becomes:

$$T_{2(Q)} = 76/(2 - Q_2)\ \text{minutes.} \qquad (1.20)$$

The overall objective with the algebraic method is to obtain from equations (1.18) to (1.20) a function which isolates one of the unknown variables (i.e. Q_1 or Q_2 or $T_{1(Q)}$ or $T_{2(Q)}$). Equation (1.18) can also be expressed as:

$$Q_1 + Q_2 = 120/T_{1(Q)} \qquad (1.21)$$

or as

$$Q_1 + Q_2 = 120/T_{2(Q)}. \qquad (1.22)$$

There is a choice of proceeding with either equation because for an equal travel time assignment:

$$T_{1(Q)} = T_{2(Q)}.$$

Equation (1.22) is chosen because the algebra is simpler. Substituting equation (1.20) into equation (1.22) and simplifying:

$$Q_1 = 3.158 - 2.579\,Q_2 \qquad (1.23)$$

$$Q_2 = 1.224 - 0.388\,Q_1 \qquad (1.24)$$

For an equal travel time assignment, equations (1.19) and (1.20) are equal. Substituting the value of Q_2 from equation (1.24) into equations (1.19) and (1.20):

$$\frac{24\,(3 - 0.7\,Q_1)}{3 - Q_1} = \frac{76}{2 - (1.224 - 0.388\,Q_1)}$$

Multiplying out the brackets, and collecting the terms together

$$-6.516\,Q_1^2 + 90.892\,Q_1 - 176.16 = 0 \qquad (1.25)$$

which is a quadratic equation of the general form:

$$a\,Q_1^2 + b\,Q_1 + c = 0$$

whose roots are solved in the standard way from:

$$Q_1 = \frac{-b \pm (b^2 - 4ac)^{1/2}}{2a}$$

either $Q_1 = 2.261$ or $Q_1 = 11.69$. The practical solution is $Q_1 = 2.261$ because the other solution gives a negative flow on route 2 which is mathematically outside the relevant boundary conditions of the problem.

From equation (1.24), $Q_2 = 0.348$; from equation (1.21) $T_1 = 46.0$; and from equation (1.22) $T_2 = 46.0$. The total amount of traffic from zone 1 to zone 2 is 2,609 vehicles per hour, with 2,261 vehicles per hour on route 1 and 348 vehicles per hour on route 2. Inter-zonal travel time is 46 minutes.

1.4.2 Transport Planning

The system equations are used to forecast the traffic implications of changes to the characteristics of transport supply. Consider, for example, two different plans: (a) close route 2 to through traffic, or (b) traffic engineering measures to improve the level of service on route 1. Students are encouraged to solve these problems by graphical and algebraic methods.

(1) Street closure of route 2: the closure of route 2 to through traffic reduces the capacity of the transport corridor and the plot of the flow-dependent travel times for route 1 is now the flow-dependent travel times for the transport corridor. Plotting the original inter-zonal traffic demand, the new equilibrium is a traffic flow of 2,370 vehicles per hour and a travel time of 51 minutes. The algebraic solution is simplified considerably because route 2 is eliminated from the analysis.

(2) Traffic engineering measures: plans to co-ordinate traffic signals along route 1 improve traffic flow, which is modelled by changing the level of service parameter from $\lambda = 0.3$ to $\lambda = 0.1$. The traffic flow-dependent travel times for route 1 are now calculated from:

$$T_{1(Q)} = 24\,(3{,}000 - 0.9\,Q_1)/(3{,}000 - Q_1)\ \text{minutes.}$$

The traffic flow-dependent travel times on route 2 remain unchanged. The new equilibrium solution is a total traffic of 2,850 vehicles per hour, with 2,650 vehicles per hour on route 1 and 200 vehicles per hour on route 2. Travel time is 42 minutes. The algebraic approach is identical to the problem solved in section 1.4.1 except for a different equal travel time assignment.

1.4.3 Land-use–Transport Planning

A typical problem in long-term planning is when the city is expected to grow larger. Two situations are examined: land-use growth in zones 1 and 2, but with the characteristics of the transport network remaining unchanged; the same land-use growth, but with an urban motorway planned to link the zone centroids. Zone 1 is expected to accommodate a total of 40,000 people, and zone 2 is expected to provide a total of 12,000 job opportunities. An increase in land-use activity means that traffic generation increases. From equation (1.11) future trip production of zone 1 is:

$$Q_{p1} = 0.4 \times 40,000 = 16,000 \text{ vehicles per hour.}$$

From equation (1.12) future trip attraction of zone 2 is:

$$Q_{a2} = 1 \times 12,000 = 12,000 \text{ vehicles per hour.}$$

From equation (1.13), the future inter-zonal traffic pattern is:

$$Q_{12} = 192,000/T_{12} \text{ vehicles per hour.}$$

This illustrative example shows a weakness of the very simple model used. Q_{12} obviously increases by a factor $\dfrac{40,000}{30,000} \times \dfrac{12,000}{10,000}$. If production had doubled and attraction had doubled, Q_{12} would have quadrupled. This contradicts common sense, but is resolved in a more refined version of the model to be presented in Chapter 3. This function is evaluated for a range of travel times and the results are plotted as in Figure 1.5 (the transport supply characteristics remain unchanged). The new equilibrium solution is 3,225 vehicles per hour, with 2,500 vehicles per hour on route 1 and 725 vehicles per hour on route 2. Travel times are 60 minutes. The algebraic approach is similar to section 1.4.1 except that equation (1.22) is replaced by:

$$Q_1 + Q_2 = 192/T_{2(Q)}.$$

Assuming the same growth in land-use activity, the introduction of a proposed motorway is likely to lower travel times and induce more trips to be made. The transport characteristics of the proposed motor-way (which is not very direct between the two zones) are: a saturation flow of 4,000 vehicles per hour; a level of service parameter 0.05; a zero-flow travel time of 18 minutes; and a length of 24 kilometres.

Figure 1.6 plots separately the traffic flow-dependent travel times for the three routes, and combines them into an equivalent traffic flow-dependent travel time function for the transport corridor. The equilibrium solution is 5,570 vehicles per hour from zone 1 to zone 2. Travel time is 34.5 minutes. The assignment of traffic is 1,780 vehicles per hour on route 1, zero on route 2, and 3,790 vehicles per hour on route 3 – the motorway.

An algebraic approach is not recommended here because it involves solving a cubic equation in terms of Q_3. The roots of a cubic equation may be solved by Cardan's method (Tranter, 1957, pp. 131–3), but the calculations are tedious.

1.4.4 A Summary

Changes to land use and transport alter the amount of traffic and the allocation of this traffic to the various parts of the transport network.

Table 1.2: Land-use–Transport Performance Measures

Alternative Situation	Accessibility (Jobs/Minute)	Travel Time (Minutes)	Transport Output (Vehicle Hours)
Existing situation	217	46	2,000
Street closure	196	51	2,015
Traffic engineering	238	42	1,995
Land-use growth	200	60	3,225
Growth plus motorway	348	34.5	3,203

Table 1.2 summarises some land-use–transport performance measures of each situation, such as accessibility to employment (calculated from equation (1.10)), inter-zonal travel time and transport output (total system travel time). The average rate at which jobs are reached in the existing situation is 217 jobs per minute. Street closures increase travel time so accessibility to employment is lower at 196 jobs per minute, whereas the traffic engineering measures reduce travel time, thereby increasing accessibility to 238 jobs per minute. Although the growth in job opportunities in zone 2 from 10,000 to 12,000 normally would improve accessibility to employment, the extra traffic and increased

Figure 1.6: Graphical Solution to Land-use—Transport Interaction Problem—Motorway Construction

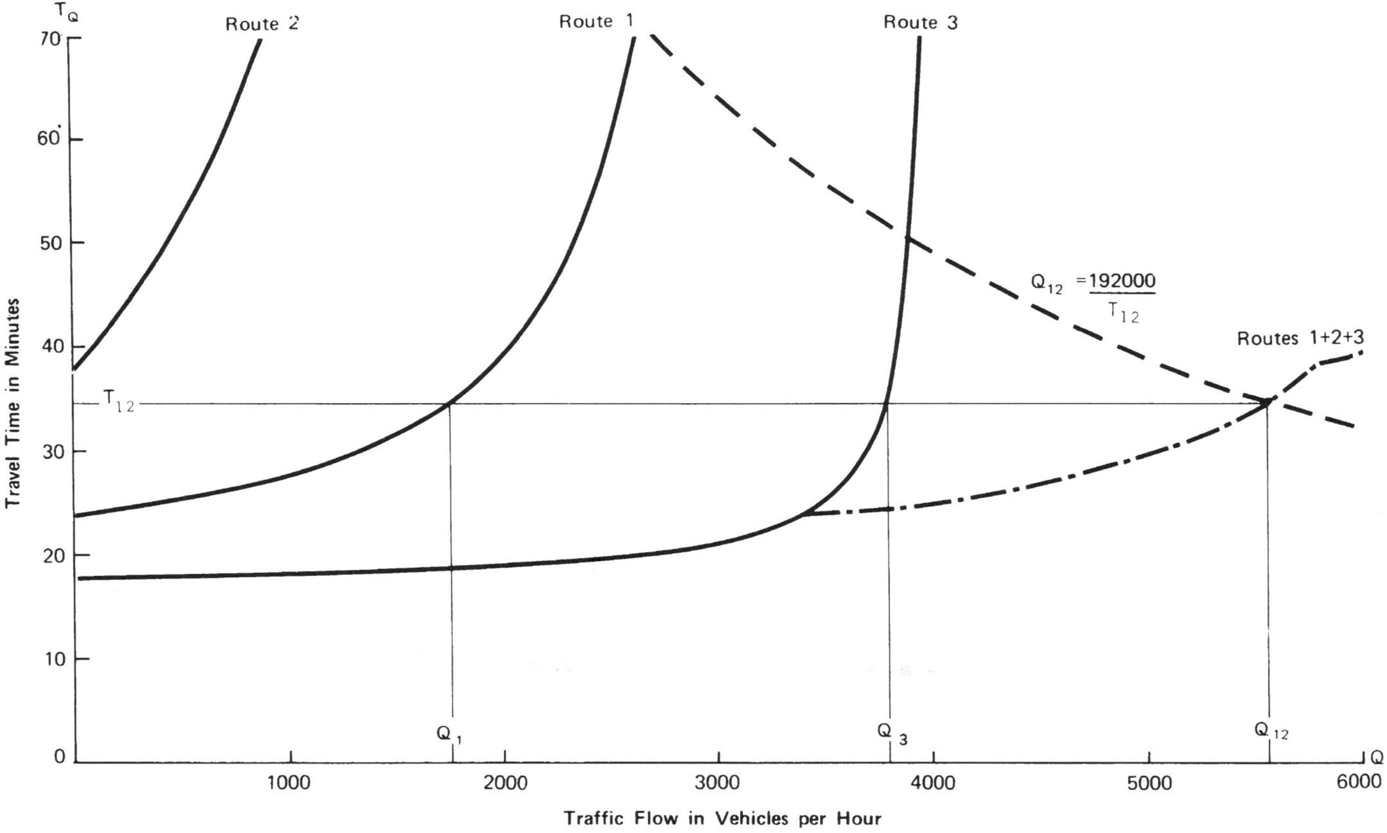

travel times result in a low level of accessibility of 200 jobs per minute. However, the construction of a motorway dramatically increases accessibility to 348 jobs per minute.

In this example, insufficient information on the relevant costs and benefits is available to allow any assessment to be made of the alternative plans. Generally, plans are beneficial if there are travel time savings to road-users, and to those people induced to travel. One measure of road-user benefit is the change in consumer surplus, as explained in section 4.6.1. The traffic engineering scheme gives a total benefit to users of 182 hours of travel saved, whereas the street closure alternative produces user disbenefits of 208 hours of extra travel. Total user benefits of 1,869 hours saved are calculated for the motorway, compared with the situation of growth in land-use activity *without* any transport improvements.

1.5 Planning the Land-use – Transport System

For planning to be effective, there is a necessity to *understand* how an urban area 'works' in terms of land use, traffic and transport *before attempting to devise solutions.* The preparation of alternative land-use and transport plans must be based therefore on a sound understanding of the way that the urban area functions, how it might evolve over time, if left to develop alone, or most importantly, how it might react to different policies.

An important aspect of the systems planning approach is to predict what would happen if there was no forward planning—sometimes referred to as the 'do-nothing' solution. The consequences of doing nothing are well worth investigating because what is likely to happen in the ordinary course of events provides a yardstick against which to assess the advantages or otherwise of any deliberate planned intervention involving changes to land use and transport.

Following the systems approach, the way to proceed with such investigations is to use equations or quantitative relationships which govern the present behaviour of the system to predict the future behaviour of the system. The traffic implications of *alternative* land-use and transport plans are calculated with the aid of the systems equation by substituting the future, anticipated values of the land-use and transport variables, which are assumed to be under some degree of control by the planner.

When alternative plans are proposed, there is the difficult problem of determining which is the best course of action to follow. The systems approach suggests that plans must be evaluated with the original goals and objectives of the study in mind, but this is controversial because plans can be judged from different, and often competing, standpoints. In principle, it is a matter of measuring and weighing up the relevant costs and benefits to all sections of the community likely to be affected. The direct costs are related to constructing new roads or railways, widening old roads, installing traffic control devices, or providing public transport services; the indirect, or more intangible costs, are related to the adverse social and environmental effects of transport. The benefits accrue mainly to transport users in the form of lower travel times and costs, and improved accessibility.

1.6 Summary

This chapter has introduced the methodology of land-use–transport planning, and has argued that the systems approach provides a convenient framework to organise the component activities of the planning process. The main steps in this planning process are the formulation of goals and objectives, data collection, systems modelling, forecasting and the evaluation of alternative plans.

Emphasis has been given in this chapter to definitions of land use, traffic and transport and to rudimentary explanations of the way that the land-use–transport system works. This system is conceptualised as a transport network connecting land-use zones, with the theoretical interactions being accessibility, traffic generation, the spatial pattern of traffic, the choice of transport mode and route and traffic on the transport network. Each concept was described and represented by a systems model. The worked example was designed to show how system models can be used to calculate the traffic implication of planned changes to land use or to transport.

The practical applications of the systems approach to urban transport planning are described in detail in Part Two of this book. The next three chapters build upon the fundamentals: Chapter 2 elaborates on data collection and the analysis of transport supply; Chapter 3 expands on the analysis of urban travel demand; and Chapter 4 explains forecasting procedures and plan evaluation methods.

2 THE ANALYSIS OF TRANSPORT SUPPLY

Public debate about urban transport often focuses upon the appropriate roles of public and private transport, and how many resources should be used in transport. A more informed debate is possible only when the location, extent and operational performance of transport facilities under prevailing traffic conditions are studied thoroughly. Such an analysis of transport supply allows the major bottlenecks and potential problem areas to be pin-pointed, and this, in turn, leads to the formulation of plans for remedial action.

This chapter discusses the collection of transport data and presents some general results based on the analysis of these data. Transport capacity is defined and ways of measuring the vehicle and passenger capacities of public transport modes, roads and intersections are described. Transport performance is explained in terms of journey speeds, delays, travel times, 'generalised time' and 'generalised cost'. Systems modelling involves both the application of graph theory to the construction of abstract networks and the specification of mathematical functions to represent the interaction between traffic and transport.

2.1 Data-collection Procedures

In preparing a highway inventory suitable for long-term planning, field-work is usually unnecessary because road authorities keep records or have most of the following information on engineering drawings: right-of-way widths; pavement widths; number of traffic lanes; surface type and condition; and capacity. Information upon which to base short-term road improvements needs to be up to date and so field-work might be required. The inventory includes critical features of road geometry, the location of traffic signs, signals and markings, and any relevant traffic laws or regulations. Parking facilities are considered in Chapter 9.

Public transport inventories include both identification of routes and the enumeration of the vehicles operating along those routes. As most of the information is available from public transport operators, field-work is unnecessary. Consultations can establish the railway rolling stock or bus characteristics such as numbers, vehicle make, model, age and type, dimensions, seating and total capacity, acceleration and

44

deceleration performance, and maximum speed. However, scheduled services run by different companies along common routes are best checked or determined from surveys.

Special surveys measure the operational characteristics of transport—speeds, journey times, delays and volumes. Speed is a common word, though it is often imprecisely defined. Correctly, it is the instantaneous speed of a vehicle at a specified location, and can be measured by radar speedmeters or calculated from photographs taken by a time-lapse camera. Field data are collected on urban roads and the average speed of vehicles calculated (equations (2.8) and (2.9)).

Under urban driving conditions, speeds fluctuate along routes so spot speeds are less informative than overall journey travel times (and hence average running speeds). The 'moving observer method' is a useful field survey to determine travel times and traffic volumes on roads (Wardrop and Charlesworth, 1954). Reliable data for main roads in an urban network can be obtained in a few weeks by a small survey team. Minimum requirements are six runs in each direction along a route (Cleveland, 1976, p. 435), but an additional four to six runs are made if large variations are obtained (O'Flaherty, 1974, p. 88). The formulae for estimating average travel times and traffic flow are explained in the next section (equations (2.10) and (2.11)).

Transport planners are especially concerned with operational delay, and field studies identify the locations where traffic conditions are unsatisfactory, measure the extent of this delay, and provide explanations. Delay is of two kinds: (a) *fixed* delay which occurs at intersections, traffic signals, railway crossings and stop signs, irrespective of the presence or absence of other traffic; and (b) *operational* delay, which is caused entirely by other traffic—vehicles blocking the roads and footpaths, vehicles parking and unparking and pedestrian-vehicle conflicts.

Similar-styled surveys are appropriate for public transport operations. An impression of public transport schedules is obtained from published time-tables, but in congested urban areas it is best to check these independently by observers riding on buses and recording the data as in the 'moving observer survey'. Additional aspects of *operational* delay to note are passenger boarding and alighting times at bus stops and stations.

2.2 Transport Capacity and Performance

When explaining the capacity and performance of different transport modes, it is useful to distinguish (a) the design features within the legally established right-of-ways and (b) the vehicles carrying the people or goods over the transport system. The various transport elements— roads, footpaths, railway tracks—and the various terminal facilities—car parks, loading docks or passenger ticket counters—all have a characteristic saturation flow or capacity. The saturation flow rate of a traffic element may be described by the distribution of the time intervals between successive units (vehicles or pedestrians) passing a reference point, providing a queue exists upstream of the reference point (Blunden, 1971, p. 161). Saturation flow can be only measured at a *point*, but the definition is commonly extended to cover the capacity of a transport link:

$$Q_{max} = 3{,}600/\bar{T} \qquad (2.1)$$

where

Q_{max} = saturation flow in units per hour;
$\bar{T}$ = mean time interval in seconds between successive units.

This time interval corresponds to the mean headway between vehicles in the case of roads or railways, or to the mean service time in the case of 'barrier' elements where the person or vehicle actually stops (ticket counters, toll booths, or parking boom gates).

Passenger capacity of those transport elements which involve vehicular conveyance is simply:

$$P_{max} = Q_{max} \cdot L \qquad (2.2)$$

where

P_{max} = passenger capacity per hour;
Q_{max} = saturation flow rate in vehicles per hour; and
L = load-carrying capacity of the individual vehicles in persons (or tonnes).

The carrying capacity of urban passenger transport depends on their dimensions, the number of seats, the amount of standing space and the socially tolerated packing density during rush-hour conditions.

2.2.1 Capacities of Public Transport

The operation of trains at safe headways is based on a wealth of practical experience. Rice (1977), extending the work of Lang and Soberman (1964), has derived equations for the minimum safe time-headway between trains and the approach speed of manually controlled trains. The point at which headways are measured is the outer home signal protecting a station berth. One equation is used if the maximum speed of the train leaving the station is *not* reached in the specified run-out distance:

$$\bar{T} = T_r + V_m\left(\frac{1}{B} + \frac{1}{2B_e}\right) + \frac{L}{V_m} + T_d + \{2(D+L)/A\}^{\frac{1}{2}} \qquad (2.3)$$

Another is used if the maximum speed *is* reached in the run-out distance:

$$\bar{T} = T_r + V_m\left(\frac{1}{B} + \frac{1}{2B_e} + \frac{1}{2A}\right) + \frac{2L}{V_m} + T_d + \frac{D}{V_m} \qquad (2.4)$$

where

$\bar{T}$ = time-headway between successive trains in seconds;
T_r = reaction time in seconds for driver and signalling;
T_d = dwell-time in seconds at the station;
V_m = maximum approach speed permitted between stations (m/sec);
B = braking rate (m/sec^2);
B_e = emergency braking rate (m/sec^2);
A = acceleration rate from rest (m/sec^2);
L = length of train in metres; and
D = run-out distance in metres.

Headways get smaller as approach speeds fall and as train lengths get shorter. Theoretically, the minimum possible headway is about 80 seconds for a maximum approach speed for 8 to 10 m/sec (29 to 36 kph) with a station dwell-time of 30 seconds. In practice, this minimum is approached on only the Moscow, New York NYCTA, and Paris Metro systems, which have trains every 90 seconds. Such headways are difficult to sustain over extended periods of time, and most busy rapid rail systems operate at most 24 to 30 trains per hour (Rice, 1977, pp. 779–80).

Similar calculations of safe headways can be made for suburban electric railways, trams and automatically guided vehicles (Black *et al.*, 1975, p. 90). Bus headways are subject to the same considerations as any vehicle traffic stream—a safe headway of about two seconds—but practical headways are dictated more by loading and unloading times at bus stops.

Table 2.1 gives the range of minimum practical headways for different public transport modes and the resulting saturation flow rate of buses or trains. The headways are those in the busy fifteen minutes in the peak hour for the main direction of passenger flow with stops along the route. These figures are intended only as a rough guide, and are based on a number of important assumptions, as detailed by Quinby (1976, pp. 232-3).

Table 2.1: Practical Minimum Mean Headways for Public Transport Systems

Public Transport System	Headways (seconds)	Vehicles/Hour per lane	Trains/Hour per track
Bus, mixed traffic	40 – 60	90 – 60	–
Bus, exclusive busway	25 – 35	144 – 103	–
Tram, mixed traffic	25 – 40	–	144 – 90
Tram, own right of way	20 – 40	–	180 – 90
Tram, automatic guidance	17	–	212
Train, suburban electric	120 – 240	–	30 – 15
Rapid Transit (Metro)	90 – 120	–	40 – 30

Source: Based on Quinby, 1976, Table 6.10, p. 233; and Black *et al.*, 1975, p. 90.

The carrying capacity of a public transport vehicle depends on its dimensions and arrangement for seating and standing passengers. To calculate the maximum capacity of a vehicle the net floor area is divided into seating space and standing space. Usually the area per seated passenger varies from 0.27 to 0.53 sq m (Quinby, 1976, p. 211) and the area per standing passenger is about 0.25 sq m (London Transport allow a more generous 0.3 sq m). Under crush conditions, this can reduce to under 0.2 sq m per standing passenger, as on the West Berlin, Budapest and Moscow metro systems (Rice, 1977, p. 806).

Table 2.2: Passenger-carrying Capabilities of Public Transport Modes

Vehicle and Right of Way	Persons per Lane (or Track) per Hour	
	Seated	Total
Buses, mixed traffic	1,800 – 5,000	2,700 – 9,000
Buses, exclusive busway	3,100 – 7,900	4,600 – 14,400
Trams, mixed traffic	5,400 – 16,200	21,600 – 37,800
Trams, own right-of-way	5,400 – 16,200	21,600 – 37,800
Trams, articulated	5,400 – 16,200	21,600 – 40,500
Trains, suburban electric	22,500 – 58,500	22,500 – 108,000
Rapid Transit	12,000 – 34,000	30,000 – 132,000

Source: Based on Quinby, 1976, Table 6.14, pp. 237–8.

The potential of urban public transport to carry large numbers of people during the rush hour is explored in Table 2.2. Passenger capacity must not be confused with passenger demand. Furthermore, the existence of theoretical excess capacity on public transport does not generate its own demand (Hamer, 1976, p. 21). For example, of the 51 metro systems surveyed throughout the world, the maximum claimed passenger capacity is in Moscow with about 72,000 passengers per track per hour; only three cities—Moscow, New York and Tokyo—generate more than 50,000 passengers per track per hour (Rice, 1977, Figure 1, p. 776).

2.2.2 Capacities of Roads

When drivers are responsible for control, the saturation flow is less easy to specify because vehicle headways depend on the type of road, its width, vehicular performance, driver behaviour and driving conditions.

Table 2.3: Design Capacities of British Urban Roads

Lanes	Width (metres)	Passenger Car Units[a] per Hour by Road Type A	B	C	D
2	6.1	300 – 500[b]	800[b]	1,200[b]	–
2	7.3	600 – 750[b]	1,200[b]	1,500[b]	–
3	9.1	900 – 1,100[b]	1,600[b]	2,000[b]	–
3	10.0	1,100 – 1,300[b]	1,800[b]	2,200[b]	–
4	12.2	800 – 900	1,200	2,000	–
4	13.4	900 – 1,000	1,350	2,200	–
4	14.6	1,000 – 1,200	1,500	2,400	3,000
6	18.3	1,300 – 1,700	2,000	3,000	–
6	21.9	1,600 – 2,200	2,500	3,600	4,500

A – all-purpose streets, capacity restricted by waiting vehicles and intersections.
B – all-purpose street, 'no waiting' restrictions and high capacity intersections.
C – all-purpose road, no frontage access, no standing restrictions and negligible cross traffic.
D – urban motorway, no frontage access, grade separation.
Notes: a. Calculated by giving the following weights to vehicle types: 0.33 (pedal cycle), 0.75 (motor cycle, scooter, moped), 1.00 (motor-car, taxi, lorry less than 1.5 tonnes), 2.00 (heavy goods vehicle), and 3.00 (bus, coach, trolley bus, tram).
b. Total capacity in both directions.

Source: Based on O'Flaherty, 1974, Table 6.4, p. 225.

Under the most ideal circumstances, a mean vehicular headway of about 1.5 seconds is theoretically possible, which gives a capacity of a single traffic lane of about 2,400 passenger vehicles per hour (Robinson, 1976, p. 309), but such flows are difficult to sustain over prolonged time-periods. A more realistic capacity for a motorway with two (or more) lanes in one direction is 2,000 passenger vehicles per lane per hour, and for an urban arterial road, about 1,800 vehicles per lane per hour. Table 2.3 reproduces the current design capacities for existing and new urban roads in Britain.

2.2.3 Capacities of Intersections

Urban road capacity is often determined by the capacity of the critical points—the major intersections. Because vehicles slow down or pause, intersections are 'barriers' and their saturation flow differs from that of a finite length of road (Miller, 1968). Theoretically, saturation flow at signalised intersections is calculated from vehicle start-up and follow-up headways and from the amount of effective 'green' time. Field studies in Britain have established empirically that the maximum rate of vehicular discharge at signalised intersections depends on the width of the road approaching the intersection:

$$Q_{max} = 525 \ W \tag{2.5}$$

where

Q_{max} = saturation flow, in passenger car units per hour of signal green-time; and

W = width of approach road in metres (distance from the edge of the carriageway to the centre-line or edge of the pedestrian refuge).

If right-turning vehicles and parked vehicles are anticipated at the intersection, the approach road width in equation (2.5) is adjusted by the following empirically verified formula (O'Flaherty, 1974, p. 236):

$$R = 1.68 - \frac{0.9 \ (D - 7.62)}{G} \tag{2.6}$$

where

R = reduction in approach width in metres;

D = distance in metres between the parked car and the intersection stop line (D has a minimum value of 7.62 m); and

G = green-time in seconds at the traffic lights.

Each right-turn manoeuvre equals 1.75 passenger car units. The lost width is multiplied by 1.5 if the parked vehicle is a lorry.

2.2.4 Performance

The average running speed of a section of railway depends on train acceleration, braking and cruising speed, on the distance separating stations, and on the dwell-time in the station (Rice, 1977, p. 793):

$$\bar{V} = S \, / \, \{ \tfrac{1}{2} \, (1/A_1 + 1/B) + S/V_m + T_d \} \qquad (2.7)$$

where all terms are defined in equation (2.4), except for:

$\bar{V}$ = average running speed (m/sec);
A_1 = average acceleration rate in (m/sec^2) up to the maximum approach speed; and
S = station separation in metres.

Operational delay is the additional dwell-time taken for boarding and alighting passengers—about one to one-and-a-half seconds per passenger through double-doors.

Door-to-door travel times are calculated by adding travel time to the station, waiting time for the next train, in-vehicle time, and the time spent in getting to the ultimate destination. Wider station spacings offer higher overall speeds but reduce convenience by lengthening the average journey between homes and the station, whereas close station spacing does the opposite. This question of optimal station spacing has attracted a good deal of attention (Newell and Vuchic, 1968; White, 1976, pp. 115–17).

Speeds of motor vehicles on urban roads rarely attain their technological maximum because of legally imposed speed limits or the constraints imposed by other traffic. The time-mean speed of vehicles, established empirically from data collected by radar speedmeters, is calculated from:

$$\bar{V}_t = \left(\sum_{i=1}^{n} V_i \right) / n \qquad (2.8)$$

where

$\bar{V}_t$ = time-mean speed of vehicle;
V_i = spot speed of the ith vehicle; and
n = number of observations.

The space-mean speed, obtained from time-lapse photography, is:

$$\bar{V}_s = \bar{D} \, / \, t \qquad (2.9)$$

where
- $\bar{V}_s$ = space-mean speed of vehicles;
- $\bar{D}$ = mean distance travelled by all the observed vehicles; and
- t = constant time interval between successive camera shutter movements.

The time-mean speed of traffic is always greater than its space-mean speed, unless (and this is improbable) there is no variation in individual vehicle speeds (Wardrop, 1952).

The 'moving observer method' is a satisfactory way of establishing journey times in urban areas. Traffic flow of the surveyed stream and the mean travel time are estimated from (O'Flaherty, 1974, p. 88):

$$Q_k = 60\,(Q_1 + X)\,/\,(T_1 + T_2) \tag{2.10}$$

$$\bar{T}_k = T_2 - (X\,/\,Q_k) \tag{2.11}$$

where
- Q_k = traffic flow for a survey stream along route k, vehicles per hour;
- Q_1 = number of vehicles met when travelling against the traffic stream;
- X = the number of vehicles overtaking the test vehicle minus the number of moving vehicles overtaken by the test vehicle, in the traffic stream;
- T_1 = travel time in minutes against the traffic stream;
- T_2 = travel time in minutes with the traffic stream; and
- $\bar{T}_k$ = mean travel time in minutes along route k.

2.3 Networks and Transport Supply

In modelling transport supply, the characteristics of the different transport modes described above require substantial simplification, and therefore detail is sacrificed for computational convenience. The complex patterns of rights of way are reduced to an abstract network and the performance of vehicles (headways, speeds) are represented by travel times or 'generalised costs' over the network. The interactions between traffic and transport, especially in the case of roads, are included in the network model by using mathematical functions of traffic flow-dependent travel times (or costs).

2.3.1 Network Representations

Graph theory deals with abstract configurations consisting of lines and points, and is suitable for representing the topological properties of transport systems. Important graph theory concepts and their transport equivalents are:

(a) *Link*–imaginary straight line that represents a finite length of road, railway or bus route.

(b) *Node*–imaginary point where links intersect. Nodes represent road intersections and railway junctions; on public transport networks nodes also indicate the location of stations or bus stops.

(a) Transport Network

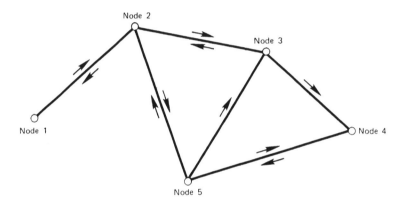

(b) Network Connectivity Matrix

FROM / TO	1	2	3	4	5
1 ➤	–	1	0	0	0
2 ➤	1	–	1	0	1
3 ➤	0	1	–	1	0
4 ➤	0	0	0	–	1
5 ➤	0	1	1	1	–

Figure 2.1: Directed Network and its Connectivity Properties

(c) *'Dummy' Link*—an additional link not corresponding with any real section of the transport system, but included to represent *either* the connection of a zone centroid with the network *or* public transport waiting time at each station or bus stop node.

The powerful aspect of graph theory exploited in transport analysis is that networks, and their connectivity properties, can be defined exclusively in numerical terms (Taaffe and Gauthier, 1973, Chapters 4 and 5; Fullerton, 1975).

Figure 2.1(a) shows the representation of one- and two-way roads as a network of six links and five nodes. Nodes are labelled with a different integer number—ordering is unimportant. The connectivity of this network is indicated by Figure 2.1(b), which is a two-dimensional array (matrix). Node labels form both row and column headings. Existence of a *direct* connection between any two nodes is shown by the element 1 in the array; absence of a direct connection between any two nodes is shown by the element 0 in the array. An alternative numerical description of connectivity is to list the node-node connecting pairs. For example, in Figure 2.1(a) the connections are: 1-2, 2-1, 2-3, 2-5, 3-2 . . . and so on.

The special representations listed below are used by transport planners, and are therefore worth noting.

(a) The dotted line in Figure 2.2(a) is a 'dummy' link, or connector between a zone centroid, which does not lie on the transport network; and the network proper.

(b) Public transport networks are more complicated than road networks because 'dummy' links are essential to represent access time to and from home (the zone centroid) to public transport and waiting time (Figure 2.2(b)). On the assumption of random passenger arrivals, waiting time for the next service is half the headway between successive vehicles when services are every fifteen minutes or less, or some arbitrary waiting time if services are infrequent.

(c) For an inter-modal transfer from one public transport service to another the 'dummy' links include walking time and waiting time. Figure 2.2(c) indicates a transfer from a bus stop to a nearby railway station.

Figure 2.3 is a map of the street pattern around the University of New South Wales campus in Sydney, with a transport network of

(a) 'Dummy' Links

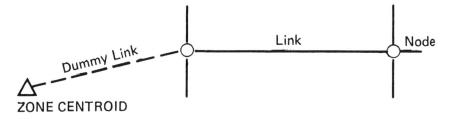

(b)Access and Waiting Time

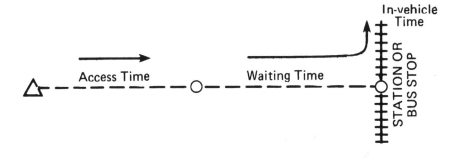

(c) Intra-modal Transfer

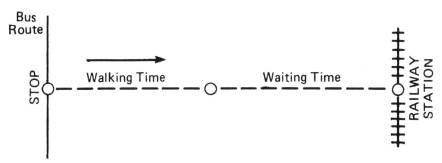

Figure 2.2: Public Transport Network Representation

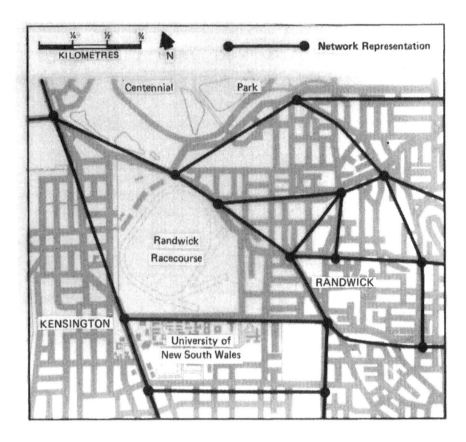

Figure 2.3: Transport Network Representation of Roads in Randwick, Sydney

links and nodes (unlabelled) superimposed. The network is a simplification because only the main traffic thoroughfares are included, and many of the road alignments are straightened. Of course, more streets could be included, although computational costs increase; this is often necessary in transport system management studies (Chapter 8) or in detailed local area planning exercises (Chapter 9).

2.3.2 Link Transport Impedances

The only feasible way of representing the complex operational characteristics of different transport modes is to construct mode-specific networks and to specify the average link travel time or cost as the measure of difficulty of getting along each link. Many of the simple worked examples in this book use link travel times because the concept is straightforward, but current UK practice involves the specification of 'generalised cost' for each link.

'Generalised cost' involves three terms: money (M), travel time (T) and the value of time (v). Expressed in monetary units, 'generalised cost' is:

$$G_c = M + v T \qquad (2.12)$$

and, expressed in units of 'equivalent' time, is:

$$G_t = T + M / v \qquad (2.13)$$

where

G_c = 'generalised cost' in monetary units; and
G_t = 'generalised time' in 'equivalent' time units.

Advice from the UK Department of Transport is to use 'generalised time' in making traffic forecasts because incomes might change, and to use 'generalised cost' at the transport plan evaluation stage (Goodwin, 1978, Table I, pp. 284–5).

The 'generalised cost' of a public transport journey is:

$$G_c = \phi D + v T_a + v T_w + v T_v + \delta \qquad (2.14)$$

where

G_c = 'generalised cost' of a journey by public transport;
D = journey distance;
T_a = travel time to get to and from public transport in minutes;
T_w = waiting time in minutes;
T_v = in-vehicle travel time in minutes;
ϕ = fare, per unit of distance;
v = monetary value of time, cents/minute; and
δ = an optional 'penalty' parameter to reflect intangible costs.

Walking and waiting time is usually counted as double, thereby reflecting travellers' perceptions of inconvenient walks or lengthy waits (Jones, 1977, p. 40). As noted by Bonsall (1976), fare structures may be more complex then the simple linear structure in equation (2.14) and suggestions are made for incorporating stepped and complex functions.

Similarly, the 'generalised cost' of a journey by motor car is:

$$G_c = \psi D + v T_v + C \qquad (2.15)$$

where terms are defined above, except for:

G_c = 'generalised cost' of a journey by car;

ψ = vehicle operating costs per unit distance; and

C = costs of parking (or tolls).

Standard values for the parameters in equations (2.14) and (2.15) are recommended unless reliable local data are available (McIntosh and Quarmby, 1970). The monetary value of time is the traveller's hourly wage rate plus any overheads for business travel, although this is challenged (Hensher, 1977a), and 20 per cent of the wage rate for all other journey purposes. The method of estimating the value of travel time is described elsewhere (Beesley, 1973, pp. 151–86; Stopher and Meyburg, 1976, Chapter 4).

For example, consider the following simple calculations of the 'generalised cost' of a bus journey over a route distance of 2 km, at an average running speed of 15 kph. Bus headways are 10 minutes, so the estimated waiting time is 5 minutes, and the actual walking time is 2 minutes at both ends. Out-of-vehicle time 'counts' double. Assume the following parameters: ϕ = 5 cents per km, ν = $1.20 per hour (2 cents per minute), and δ = 6 cents. Therefore

$$G_c = (5 \times 2) + 2 \times (2 \times 2) + 2 \times (2 \times 5) + (2 \times 8) + 6 = 60 \text{ cents.}$$

The 'generalised time' of this journey in equivalent time units is 30 minutes.

2.3.3 Traffic Flow-dependent Travel Times

Link impedances are influenced by the amount of traffic using the transport facility, as explained in the previous chapter, and the modelling of link transport impedances attempts to include this relationship. Although travel times along the same route are often highly variable, even for similar road traffic flow conditions, modelling assumes an overall average traffic flow-dependent function. Desirable theoretical properties of traffic flow-dependent travel times (or cost) functions are listed by Blunden (1971, pp. 81–2):

(a) the intercept on the travel time axis is clearly defined as the 'zero-flow' travel time (or the reciprocal of the mean of the free speed distribution);

(b) the curve is monotonically increasing, initially at a gentle gradient, at least until saturation flow is imminent; and

(c) the curve becomes asymptotic to the saturation flow ordinate (capacity) under 'steady-state' system conditions.

Many mathematical functions are suitable but the following are the most likely to be encountered in transport studies. The first function fulfils all three theoretical requirements (Davidson, 1966); and was used in section 1.4:

$$T_Q = T_0 \frac{1 - (1 - \lambda)\, Q/Q_{max}}{1 - Q/Q_{max}} \qquad (2.16)$$

where

T_Q = travel time at traffic flow Q;
T_0 = 'zero-flow' travel time;
Q = traffic flow, vehicles per hour;
Q_{max} = saturation flow, vehicles per hour; and
λ = level of service parameter.

The level of service parameter is related to the type of road, road widths, the frequency of traffic signals and pedestrian crossings and parked vehicles (Menon *et al.*, 1974). In the absence of authentic data, Blunden (1971, p. 84) suggests λ values of from 0 to 0.2 for motorways, from 0.4 to 0.6 for urban arterials, and from 1 to 1.5 for collector roads. A method for estimating zero-flow travel times, saturation flows and the level of service parameter is described by Taylor (1977). Akcelik (1978), in an attempt to avoid the computational problems associated with infinite travel times when traffic flow equals saturation flow, draws a tangent at the point on the curve which represents 'critical' traffic flow and that part of the tangent connecting the point with the vertical saturation flow ordinate is the modified traffic flow-dependent travel time relationship.

The traffic flow-dependent travel time relationship widely used in US transport studies is the general polynomial function:

$$T_Q = T_0 \{1 + \alpha(Q / Q_{max})^n\} \qquad (2.17)$$

where

T_Q = travel time at traffic flow Q;
T_0 = 'zero-flow' travel time;
Q = traffic flow, vehicles per hour;
Q_{max} = 'practical capacity' which is defined as three-quarters of saturation flow; and
α, η = parameters.

This function has been verified empirically for North American driving

conditions. 'Zero-flow' travel time is estimated by taking field observations of traffic and travel time along roads full to 'practical capacity'
and multiplying this average travel time by 0.87. The standard parameters
in the Bureau of Public Roads traffic assignment programme are $\alpha =$
0.15 and $\eta = 4$, but the computer programme released more recently
by the US Department of Transportation contains parameter values of
$\alpha = 0.474$ and $\eta = 4$.

British transport studies have preferred to develop empirical speed-
flow relationships, following investigations by the Road Research
Laboratory (RRL, 1965, pp. 113-14) into the relationship between
the running speed of vehicles (excluding intersection delay) and traffic
flow. For new roads in central urban areas (O'Flaherty, 1974, p. 206):

$$V_Q = 49.9 - 0.163 \, (Q + 430) / (W - R) \qquad (2.18)$$

This holds for $16 < V_Q < 38$ kph. For other urban roads, and for
$32 < V_Q < 56$ kph:

$$V_Q = 67.6 - 0.123 \, (Q + 1,000) / (W - R) \qquad (2.19)$$

where

V_Q = speed in kph at traffic flow Q;
Q = traffic flow in vehicles per hour;
W = carriageway width in metres; and
R = reduction in carriageway width in metres by, for example,
     parked vehicles.

2.4 Summary

For analytical purposes transport supply is represented as a network of
nodes and links, with transport impedance associated with each link.
This difficulty in getting around on transport is measured by travel
time, cost, or some weighted combination of these, such as 'generalised
cost' or 'generalised time'. Transport impedances on links of a road
network vary by the amount of traffic carried and this is modelled by
a mathematical function that specifies speed-flow relationships or
traffic flow-dependent travel time relationships.

Much of what follows in the next chapter on the analysis of travel
demand assumes a familiarity with the analytical representation of
transport supply. The section on traffic assignment (3.5), in particular,
should be read with this chapter in mind. The representation of future
transport supply in transport plans is explained in section 4.5.

3 THE ANALYSIS OF TRAVEL DEMAND

This chapter introduces urban travel demand models. Data-collection procedures, model specification, model calibration techniques and model verification are discussed. Aggregate models based on zonal data and behavioral models based on individual data are described. Simple numerical examples are given to emphasise the model structures and the calibration procedures.

Travel-demand modelling is a large topic and so selectivity is essential. The models included here are either commonly encountered in transport analysis or are applied in planning practice, as illustrated in Part Two of this book. The broad class of econometric models (Stopher and Meyburg, 1975, Chapter 15), and optimisation models (Black and Blunden, 1977) are excluded. Alternative models of traffic distribution to those described are growth factor methods (Martin *et al.,* 1961, pp. 127-38), linear programming methods (Blunden, 1971, Chapter 4) and the intervening opportunities model (Ruiter, 1967).

3.1 Data-collection Procedures

Information-gathering and the coding of data are important parts of urban transport planning and this aspect of the systems approach absorbs, typically, from one-half to two-thirds of the total budget (Hutchinson, 1974, p. 385). The aims and objectives of a transport study dictate the details and amount of information to be sought, but all collect data on land-use and travel patterns. Land-use surveys establish the allocation of land for different purposes in the study area and the intensity of social and economic activities taking place on the land. Surveys of households and firms provide quantitative information on the movement of people and goods (Bruton, 1975, Chapter 2).

3.1.1 Surveys

Urban planning authorities conduct a range of land-use surveys (Chapin, 1976, Chapter 7), but it is uncommon for the data collected to be made available in a form suitable for further analysis. Census data are more useful but the survey dates do not always coincide conveniently with the timing of transport studies. Consequently, urban transport studies include: (1) home interview surveys, (2) commercial vehicle surveys, (3) roadside interviews and (4) public transport surveys.

(1) Home interview survey: ideally information is sought on house-hold characteristics and all residents' travel patterns, but this is too expensive and data are collected for a sample of households throughout the study area. Sample size depends on city size and on the level of statistical accuracy required, but sampling theory and experience enable the US Department of Transportation to stipulate recommended and minimum sample rates when conducting home interview surveys in urban areas of differing populations (Table 3.1). The usual method of drawing a sample is to take a systematic sample from either the electoral rolls, the property valuation lists, or from consumer lists supplied by the electricity authority.

Table 3.1: **Recommended and Minimum Sample Size for Dwelling Units in a Home Interview Survey**

Study Area Population	Sample Size	
	Minimum	Recommended
Under 50,000	1 in 10	1 in 5
50,000 – 150,000	1 in 20	1 in 8
150,000 – 300,000	1 in 35	1 in 10
300,000 – 500,000	1 in 50	1 in 15
500,000 – 1 million	1 in 70	1 in 20
Over 1 million	1 in 100	1 in 25

Source: Based on Bruton, Table 1, p. 53.

The usual procedure is for an interviewer to call on a household on a scheduled date and to leave a copy of the home interview question-naire. Examples of the layout of questionnaires are given in Bruton (1975, Figure 6, pp. 56-7) and in Stopher and Meyburg (1975, Figure 5-8, pp. 82-5). This questionnaire is broadly divided into: (a) general household characteristics—number of residents, vehicles owned or garaged, income, dwelling type; (b) characteristics about family members —occupation, sex, age; and (c) individual travel information about journeys actually made during the stipulated survey period—trip origin and destination, purpose, land-use at origin and destination, travel time and transport mode.

The questionnaire should make it very clear what is meant by a 'trip'. Although the trip is the basic unit of traffic measurement, study definitions differ, as noted by Black and Salter (1975a). A trip is a purposeful journey by any transport mode (including walking) from one place to another, usually covering a distance of a street-block, or more. For most journeys this definition is satisfactory, but multipurpose or multi-modal trips are a source of potential confusion that can be avoided only with a careful enumeration by the traveller of each part of a journey.

(2) Commercial vehicle survey: a similarly styled survey of non-residential land uses could be designed to collect information on goods movements, but transport resources are rarely allocated to such an ambitious project. Instead, urban freight flows are usually measured indirectly from a commercial vehicle survey. There are often difficulties in establishing the 'population' of commercial vehicles from which to draw a random sample because (a) it is impossible to distinguish all station-wagons, passenger utilities and vans registered privately but used for business; (b) some vehicles have registration addresses within the study area but are used elsewhere, or they are registered elsewhere but operate within the study area; and (c) some vehicles are temporarily out of service or have been scrapped. Contacting firms and businesses directly is time-consuming and inefficient because not all firms operate their own vehicle fleets.

The questionnaire is usually divided into: (a) characteristics of the firm; and (b) journeys made by each sampled vehicle, including the type and weight of the truck and the commodity carried.

(3) Roadside survey: roadside surveys obtain information on the origin and destination of through traffic, or the destination in the study area of external traffic. Drivers are asked questions about the origin and destination of their journey, vehicle type, occupancy and journey purpose. Alternatively, a pre-stamped postcard questionnaire is given to drivers, but response rates are often 50 per cent or lower. The sample size varies with the traffic volume, because an interviewer can cope with about eighty vehicles per hour in a properly organised survey (Bruton, 1975, p. 61).Automatic or manual traffic counts are taken at the same time to determine the sampling rate, and to check independently the accuracy of vehicular trips reported in the home interview survey.

(4) Public transport passenger survey: a survey conducted on public transport is usually needed to collect information on journeys which start outside and have a destination within the study area. Questionnaires

are issued at stations or bus stops outside the study area to all inbound passengers, who are asked to fill out the questionnaire, delivering it to a convenient collection point or posting it to the survey address. The questions include journey origin and destination, purpose, transport mode prior to boarding, and transport mode after alighting (Watson, 1978).

3.1.2 Data Assembly

Data collected from sample surveys are multiplied by an appropriate expansion factor (Bruton, 1975, pp. 79-81) to represent the whole population. Subsequent analysis may take place using the original data on individuals (disaggregated analysis) or the data grouped in some convenient way (aggregate analysis). As the purpose of transport analysis is to investigate the adequacy of transport supply in different locations, it is sensible to aggregate the data into spatial units, or zones.

The following factors are relevant in the design of a zoning system for a study area.

(a) The zone should contain predominantly one land use.
(b) Characteristics of the activities within the zone should be as homogeneous as possible (Cliff *et al.*, 1975, pp. 17-19).
(c) A compromise between a small zonal area that gives locational precision and a large zonal area that gives a sufficient sample of households or firms to produce statistically reliable traffic estimates (BPR, 1967, p. 55). To give an indication of size, residential zones, in small towns, contain from 1,000 to 3,000 people, and in large cities from 5,000 to 10,000 people (Lane *et al.*, 1971, p. 33; Wells, 1975, p. 71).
(d) The zone system should be compatible with the boundaries of census enumeration districts.
(e) The zone boundaries follow, where possible, major roads, railways, canals, or other physical barriers to movement.

The spatial aggregation of data is an important theoretical consideration because the results of any analysis are not entirely independent of the delineation of zones (Openshaw, 1978) and different spatial aggregations may lead to different, and sometimes contradictory, conclusions (Batty and Sammons, 1978).

The principles of modelling are introduced in the next four sections with the highly aggregated models of total traffic. However, the analysis may be stratified by trip purpose, time of day or population subgroup to suit the requirements of the analyst. In transport planning practice, daily traffic is stratified by trip purpose, and usually into peak and off-peak travel. In towns with populations less than 100,000, traffic is divided into home-based work trips, home-based other trips, other non-home-based trips and commercial vehicle trips. In larger cities, the recommended stratification (BPR, 1965, p. 111-18) for residential traffic is: home-based work; home-based school; home-based shopping; home-based personal business; home-based social/recreation; and non-home-based trips. Other traffic categories are light and heavy commercial vehicles, and taxis.

3.2 Traffic Generation

The purpose of traffic generation modelling is to develop equations that estimate the amount of traffic generated by zones. Because a trip has two ends, separate analyses are made for the traffic produced by a zone, and the traffic attracted to a zone, but considerable care must be taken to dimension precisely the traffic units of measurement. For each zone in the study area, the surveys provide information on the observed number of trips produced and attracted, and a range of land-use and socio-economic variables. Zonal tabulations of these data are used to build models of traffic generation.

3.2.1 Two Models of Traffic Generation

In Chapter 1 it was sufficient to state theoretically that traffic is related to land-use, but systems modelling involves specification of the exact functional form of the relationship and estimation of the parameters for a set of data. The problem is one of finding those land-use variables, from an array of all land-use variables, that significantly influence the amount of traffic generation. Two approaches to this problem are explained—linear regression analysis and category analysis.

Regression analysis is a statistical method for studying how a dependent variable is related to one (or more) explanatory variables. In the simplest case, confined to two variables, the general relationship is:

$$Y = \alpha + \beta X \qquad (3.1)$$

where

Y = dependent variable;
X = explanatory (independent) variable;
α = regression constant; and
β = regression coefficient.

If equation (3.1) is used to study zonal traffic production then all variables are identified by the subscript i; if the equation is used to study zonal traffic attraction then all variables are identified by the subscript j.

Social data contain relationships inherently more complex than can be represented by the simple linear regression model. For example, a number of land-use variables simultaneously influence the amount of traffic generation. Multiple linear regression analysis is appropriate for exploring the relationship between a dependent variable and two, or more, explanatory variables. The general relationship, containing a total number of z explanatory variables, is:

$$Y = \alpha + \beta_1 X_1 + \beta_2 X_2 + \ldots \beta_z X_z \tag{3.2}$$

where

Y	=	dependent variable;
$X_1 \ldots X_z$	=	explanatory (independent) variables;
α	=	regression constant; and
$\beta_1 \ldots \beta_z$	=	partial regression coefficients.

Regression analysis is a statistical method and it is important to note four major *statistical* assumptions:

(a) The variables are measured by the surveys without any inaccuracies or error.

(b) The dependent variable is a linear function of each explanatory variable. If the relationship is non-linear then mathematical transformations of the data can be performed (see Hutchinson, 1974, Figure 2.2, p. 42).

(c) The effects of the explanatory variables on the dependent variable are additive, and there is no strong relationship (correlation) between the explanatory variables.

(d) The variance in the data about the regression 'surface' must be similar for all magnitudes of the explanatory variable.

Category analysis is an alternative model of traffic generation. Winsten (1967, p. 17) explains that 'category analysis is a very proper form of regression analysis' which does not make assumptions about the shape or form of the 'response surface' between a dependent and an explanatory variable. Category analysis is far less restrictive in its underlying assumptions because it can handle non-linear relationships between traffic generation and land-use variables.

Category analysis, developed originally in the Puget Sound Regional Transportation Study (Walker, 1968), is predominantly a model of traffic production from residential land uses. On empirical evidence, three land-use variables are *assumed* to influence household traffic generation; these are car ownership, household size and household income (Wootton and Pick, 1967, Figures 1-3, pp. 138–40). Households are cross-classified by these variables, and the mean traffic production for each cross-classification is calculated from the home interview survey. The model is applied to estimate zonal traffic by (a) multiplying the mean trip rate for each category by the number of households in each category that are located in the zone and (b) summing the amount of traffic for every household category.

3.2.2 Calibration

'Calibration' of a regression model involves estimating the numerical values for the regression constant and the regression coefficients in equation (3.1) and (3.2). The principles of calibration are introduced using a simple numerical example. It is assumed that the zonal daily trip production of residential land is a function of a single land-use variable–the total number of motor cars owned by residents. The study area is partitioned into eight zones and, the following survey data are tabulated:

Zone Number	1	2	3	4	5	6	7	8
Zonal Trip Production	500	300	1,300	200	400	1,200	900	1,000
Zonal Car Ownership	200	50	500	100	100	400	300	400

In Figure 3.1, the vertical Y-axis of the graph represents the numerical range of zonal trip productions and the X-axis represents the range of the explanatory variable–zonal car ownership. The data points indicate there is a tendency for small values of car ownership to be associated with small values of traffic, and for large values of car ownership to be associated with large values of traffic, and that overall the trend is approximately linear.

The objective of calibrating a model is to replace the empirical data in Figure 3.1 by a mathematical equation which generalises the relationship. One way of doing this is to fit the best straight line 'by eye' to the data and to determine the equation of the line. Equation (3.1) is the general form of a straight line, where β is the gradient of the line (the number of units change in the vertical axis for a one-unit change along the horizontal axis) and α is the numerical value of the point on the Y-axis where the line crosses. From Figure 3.1 an approximate

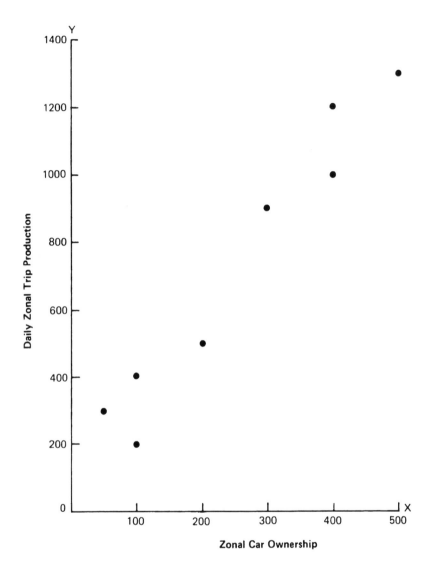

Figure 3.1: Zonal Traffic Production and Car Ownership

solution is β equals 2.5 and α equals 90. Obviously, fitting lines by eye is unreliable because different lines and different equations may be derived by different analysts.

The scientific approach—in concept, identical to the above— that fits a line to data and estimates parameters efficiently is the method of least squares. The line which 'best fits' the data is the one that minimises the sum of the squares of the vertical deviations from the best-fit line to data points. There is only one line which minimises the sum of the squares of the errors and this is the best fit, or regression, line.

The method of least squares gives rise to formulae for estimating the parameters α and β in equation (3.1):

$$\beta = \frac{n \, \Sigma \, X \, Y - \Sigma \, X \, \Sigma \, Y}{n \, \Sigma \, X^2 - (\Sigma \, X)^2} \qquad (3.3)$$

$$\alpha = (\Sigma Y)/n - \beta(\Sigma X)/n \qquad (3.4)$$

where
    n = the number of paired observations (the number of zones);
    Y = observed value of the dependent variable Y one value for each zone); and
    X = observed value of the explanatory variable X (one value for each zone).

Equations (3.3) and (3.4) can be used to verify the correct line which relates residential trip production and car ownership in Figure 3.1 is $\hat{Y} = 89.9 + 2.48 \, X$. This equation is an example of a calibrated traffic generation model, and the hat ($\hat{}$) above the Y indicates that the dependent variable in the equation provides *estimates* of traffic generation.

Conceptually, the calibration of the multiple linear regression model is similar to simple regression. There are standard formulae for estimating the parameters of equation (3.2) but it is unnecessary to state them here because manual calculations involving several explanatory variables are lengthy. Standard programmes are available for desk calculators or for digital computers (Baxter, 1976, pp. 263–7).

In constructing a category analysis model, four main steps are followed which replace calibration procedures.

(1) A definition of each possible cross-classification. For example, in the West Midlands Transportation Study six categories of household size, six categories of household income and three levels of household car ownership were specified. The categories of household size and structure were defined as: (a) no employed residents and one non-employed adult, (b) no employed resident and two (or more) non-employed adults, (c) one employed resident and one (or less) non-employed adult, (d) one employed resident and two (or more) non-employed adults, (e) two (or more) employed residents and one (or less) non-employed adult, and (f) two (or more) employed residents and two (or more) non-employed adults. The

six categories of household annual income included the following ranges: (a) less than £500, (b) £500 to £1,000, (c) £1,000 to £1,500, (d) £1,500 to £2,000, (e) £2,000 to £2,500 and (f) more than £2,500. The household car ownership groups were: (a) zero cars, (b) one car and (c) two (or more) cars.

(2) An allocation of every household into one of the cross-classifications involves sorting through the home interview data file, and pigeon-holing each household into the appropriate one of a possible 108 categories ($6 \times 6 \times 3$).

(3) A calculation of the average traffic production rate for each of the 108 categories. The average household trip rate is obtained separately for each category by dividing the total number of trips by the total number of households.

(4) An estimation of zonal trip productions. The number of households in each category is multiplied by the category traffic production rate, and the results are summed over all 108 categories.

$$\hat{Q}_{pi} = \sum_{c=1}^{108} \bar{Q}_c \cdot N_{ci} \qquad (3.5)$$

where

$\hat{Q}_{pi}$ = estimate of the zonal number of trips produced by zone i;
$\bar{Q}_c$ = mean trip-rate for category c type households; and
N_{ci} = number of category c type households located in zone i.

Table 3.2: Category Analysis Household Trip-Rates—an Example

Car Ownership Levels	Income Levels		
	Low	Medium	High
Zero	3.4[a]	3.7[a]	3.8[a]
	4.9[b]	5.0[b]	5.1[b]
One	5.2[a]	7.3[a]	8.0[a]
	6.9[b]	8.3[b]	10.2[b]
Two, or more	5.8[a]	8.1[a]	10.0[a]
	7.2[b]	11.8[b]	12.9[b]

Notes: a. Trip rate for household sizes one to three persons.
   b. Trip rate for household sizes with four or more persons.

The principles of category analysis are demonstrated with a numerical example, deliberately simplified to eighteen cross-classifications. The definition of each category variable results in two household sizes (with either 1 or 3 persons, or 4 or more persons), three income levels (low, medium and high), and three car ownership levels (zero, one, or two or more cars). The average daily trip production rates for the eighteen categories are given in Table 3.2. To avoid the representation of a three-dimensional array, the table shows nine categories of income and car ownership and uses superscript a to distinguish the traffic production rate of large households from small households (marked with superscript b).

The problem is to estimate the traffic generation for a zone which contains: 100 households with low income, no car and three persons; 200 households with low income, no car and four persons; 300 households with medium income, one car and four persons; and 50 households with high income, two cars and five persons. The calculations are:

$$(100 \times 3.4) + (200 \times 4.9) + (300 \times 8.3) + (50 \times 12.9)$$
$$= 4,455 \text{ daily trips produced per zone.}$$

3.2.3 Model Validation

Model validation investigates how accurate the estimates from a calibrated regression model are when compared with the observed data, and checks to see whether the best combination of explanatory variables is included in the model specification. This is unnecessary for the category analysis model of residential trip production because the three explanatory variables are assumed, and because there are no statistical tests that assess goodness-of-fit.

The search for the best model specification is aided by computer programmes, especially the step-wise regression method which automatically selects various combinations of explanatory variables. It is sound practice to plot each potential explanatory variable against the dependent variable on a graph and examine the general form of the relationship. The variables that enter into the final model should come as no surprise to the analyst. In fact, a good model specification and validation emerge from a mixture of common sense and interpretation of statistical tests.

(a) Plausibility: Does the regression coefficient have a positive sign for an explanatory variable that logically contributes to an increase in traffic? Does the regression coefficient have a negative sign when that variable is expected to cause a decrease in traffic?

(b) Simplicity: Keep the number of explanatory variables to a minimum.

(c) The standard statistic for the 'goodness-of-fit' of a simple linear regression model is the correlation coefficient (r), whose formula is:

$$r = \frac{n\,\Sigma X \cdot Y - \Sigma X\,\Sigma Y}{\sqrt{\{n\,\Sigma X^2 - (\Sigma X)^2\}\,\{n\,\Sigma Y^2 - (\Sigma Y)^2\}}} \tag{3.6}$$

An r value of plus one is a perfect positive correlation, an r value of minus one is a perfect negative correlation, and an r value of zero indicates no correlation. The multiple correlation coefficient is the equivalent statistic for the multiple linear regression model, and has values in the range plus one to minus one. A fuller discussion of these ideas, and other statistical tests, is given by Hutchinson (1974, pp. 37–46).

3.3 Traffic Distribution

The purpose of traffic distribution modelling is to find an equation that reproduces the intra- and inter-zonal pattern of survey traffic. Figure 3.2 identifies the relevant survey information used for building a model. The origin-destination matrix shows the amount of traffic from any zone i to any zone j (including $i=j$). The summation of traffic along each row is zonal traffic production and the summation of traffic down each column is zonal traffic attraction. The second matrix in Figure 3.2 has an identical number of rows and columns and contains survey information about intra- and inter-zonal transport impedance, measured by distance, travel time or cost (BPR, 1965, p. IV–15). The trip-length frequency distribution indicates the amount of traffic that falls within intervals of distance, time or cost. A useful summary statistic is the mean trip length.

3.3.1 Gravity Models for Traffic Distribution

The approach to building traffic distribution models is based on reasoning by analogy from the physical sciences. The proposition that human spatial behavior is governed by something equivalent to Newton's law of gravity has been extensively exploited (Carrothers, 1956). The physical law states that the force which each body in the universe exerts on another is proportional to the product of their masses and inversely proportional to the square of their distance apart. By analogy,

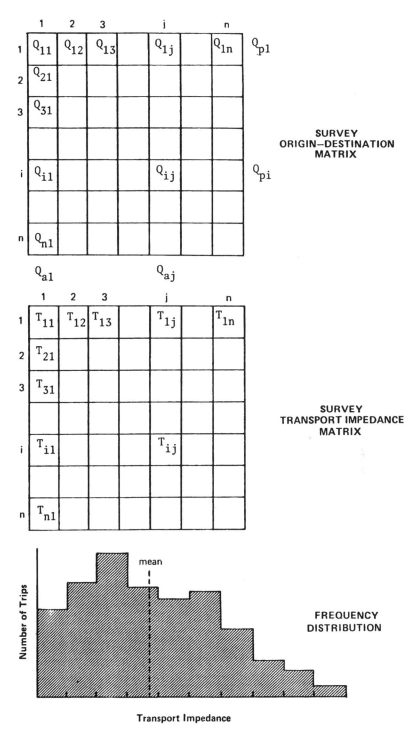

Figure 3.2: Information for Building a Gravity Model

the amount of traffic interaction between two places is directly proportional to the product of the amount of land-use intensity in both places and inversely proportional to the transport difficulty in getting from one place to the other:

$$Q_{ij} = \kappa L_{oi} L_{dj}/f(T_{ij}) \qquad (3.7)$$

where the constant κ is included to convert the somewhat arbitrary multiplication of land-use intensities and division of transport impedances into units corresponding with traffic flows.

In a major theoretical contribution, Wilson (1967, 1970, 1974) has demonstrated how the gravity model can be derived from entropy-maximising methods. The Newtonian analogy takes zonal land-use intensity as 'masses' whereas the 'entropy-maximising method works with individuals, assesses their probability of making a particular journey . . . and, essentially, obtains the interaction as a statistical average' (Wilson, 1974, p. 393).

The four traffic distribution models which preserve the structure specified by equation (3.7) differ in the formulation of the constant term, and the measures of land-use intensity. As traffic is a function of land-use, zonal measures of land-use are replaced by zonal measures of traffic production and traffic attraction:

(1) The Unconstrained Gravity Model

$$\hat{Q}_{ij} = \kappa Q_{pi} Q_{aj}/f(T_{ij}) \qquad (3.8)$$

where

$$\kappa = \sum_i \sum_j Q_{ij} \Big/ \sum_i \sum_j \hat{Q}_{ij}$$

(2) The Production-constrained Gravity Model

$$\hat{Q}_{ij} = \kappa_i Q_{pi} Q_{aj}/f(T_{ij}) \qquad (3.9)$$

where

$$\kappa_i = \{\Sigma Q_{aj}/f(T_{ij})\}^{-1}$$

(3) The Bureau of Public Roads (Production-constrained) Gravity Model

$$\hat{Q}_{ij} = \kappa_i Q_{pi} Q_{aj} F(T)$$ (3.10)

where

$$\kappa_i = \{\Sigma Q_{aj} F(T)\}^{-1}$$

$F(T) =$ empirically derived 'friction factors' which measure the average area-wide effect of the spatial separation of those $i - j$ zone pairs that are in the travel time interval, T.

(4) The Fully Constrained Gravity Model

$$\hat{Q}_{ij} = \kappa_i \kappa'_j Q_{pi} Q_{aj} / f(T_{ij})$$ (3.11)

where

$$\kappa_i = \{\Sigma \kappa'_j Q_{aj} / f(T_{ij})\}^{-1}$$

and

$$\kappa'_j = \{\Sigma \kappa_i Q_{pi} / f(T_{ij})\}^{-1}$$

The single constant term, κ, in the unconstrained model ensures that the total number of trips in the model origin-destination matrix equals the total number of trips in the survey origin-destination matrix, but when the trips in each row and column of the model matrix are added up to give the estimates of zonal traffic production and attraction, these totals do not necessarily equal the survey traffic productions and attractions. The production-constrained models contain a term for each production zone (κ_i) and this zonal variable ensures that when the trips in each row of the model matrix are added up, the estimates of zonal production equal the survey zonal traffic production. However, the column totals for each zone–the estimates of zonal traffic–do not equal the survey zonal traffic attraction. The fully constrained gravity model contains a constant for each production zone (κ_i) and for each attraction zone (κ'_j). Their product for all $i - j$ pairs ensures that both the row and column totals of the model matrix equal the row and column totals of the survey matrix.

3.3.2 Calibration

Calibration of the gravity models (equations (3.8), (3.9) and (3.11)) involves: (a) specification of the mathematical function to substitute for $f(T_{ij})$ that replicates the shape of the survey trip-length frequency distribution and (b) adjustments to the parameters of this function so that the mean trip length of the model traffic equals the mean trip length of the survey traffic.

There are a number of appropriate mathematical functions (Openshaw and Connolly, 1977) but the most common in transport analysis are:

(a) the power function, $f(T_{ij}) = T_{ij}^{\alpha}$;
(b) the exponential function, $f(T_{ij}) = \exp(\beta T_{ij})$; or
(c) Tanner's function, $f(T_{ij}) = T_{ij}^{\alpha} \cdot \exp(\beta T_{ij})$.

A rigorous calibration involves testing different functions and choosing the one that best fits the empirical shape of the survey trip length frequency distribution (Black and Salter, 1975b) but this is rarely done in practice.

Calibration involves finding the numerical value of the parameter α or β, or the two parameters α and β, which control the model mean trip length. The simplest procedure is to run the model for a range of parameter values, thereby calibrating the model by 'trial and error', but a more efficient method is to adopt a systematic search routine, such as the Fibbonaci or golden-section search algorithm (Batty, 1971) or the Newton-Raphson method (Batty and Mackie, 1972). The parameters are chosen to ensure that:

$$(\sum_i \sum_j \hat{Q}_{ij} T_{ij}) / \sum_i \sum_j \hat{Q}_{ij} = (\sum_i \sum_j Q_{ij} T_{ij}) / \sum_i \sum_j Q_{ij}$$

Low (positive) parameter values are associated with relatively *long* model mean trip lengths and *high* (positive) values are associated with relatively *short* model mean trip lengths. A more detailed technical discussion of calibration methods is found in the literature (Evans, 1971; Kirby, 1974; and Williams, I., 1977).

In the calibration of the Bureau of Public Roads gravity model (equation (3.10)) a set of empirically determined 'friction factors' replaces the mathematical function of transport impedance because of difficulties in specifying the correct mathematical function and a desire to 'simplify computational procedures' (Hansen, 1962, p. 68). Inter-zonal travel time is recommended as the measure of transport impedance (BPR, 1965, p. IV–12). The friction factors appear in the numerator of the gravity model: *large* numerical values for the friction factors are associated with relatively *short* inter-zonal travel times; *small* numercial values are associated with relatively *long* inter-zonal travel times.

Calibration involves finding the numerical values for $F(T)$ for each of the predetermined set of travel time intervals (usually five-minute intervals) by an iterative procedure. Initially, a set of friction factors is assumed. By setting $F(T)$ to unity for all travel time intervals, the first model origin-destination matrix is calculated from:

$$\hat{Q}_{ij} = Q_{pi}Q_{aj}/\sum_j Q_{aj} \qquad (3.12)$$

The number of model trips in each travel time interval is compared with the number of survey trips in the corresponding time interval, and a revised set of friction factors are calculated from:

$$F'(T) = F(T)Q(T)/\hat{Q}(T) \qquad (3.13)$$

where

$F'(T) =$ revised friction factor for travel time interval T;
$F(T) =$ previous friction factor for travel time interval T;
$Q(T) =$ number of survey trips in travel time interval T;
$\hat{Q}(T) =$ number of model trips in travel time interval T.

This process is repeated until the new friction factor, for all time intervals, differs by only a small amount from the previous iteration. 'This is a relatively arbitrary closure criterion, and may sometimes be replaced by the use of a statistical test, such as a Chi-square test or a Kolmogorov-Smirnov test' (Stopher and Meyburg, 1975, p. 144). The difference between the model and survey mean trip lengths should be within ±3 per cent.

The Bureau of Public Roads model is a production-constrained model, and the estimates of zonal traffic attraction do not equal the survey zonal traffic attractions. A procedure that eliminates these discrepancies is to compare the survey and model column totals, and to formulate a new zonal traffic attraction value from (BPR, 1965, p. IV-25):

$$\hat{Q}'_{aj} = (Q_{aj}/\hat{Q}_{aj})Q_{aj} \qquad (3.14)$$

The revised attraction values are substituted into the calibrated gravity model:

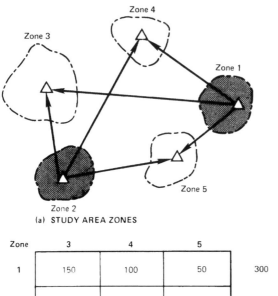

(a) STUDY AREA ZONES

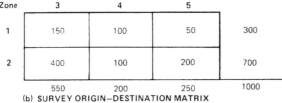

Zone	3	4	5	
1	150	100	50	300
2	400	100	200	700
	550	200	250	1000

(b) SURVEY ORIGIN–DESTINATION MATRIX

Zone	3	4	5
1	3	2	5
2	3	5	4

(c) INTER–ZONAL DISTANCE MATRIX (km)

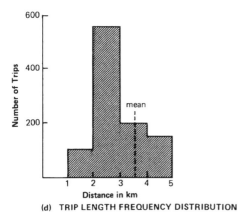

(d) TRIP LENGTH FREQUENCY DISTRIBUTION

Figure 3.3: Gravity Model Calibration Exercise

$$\hat{Q}_{ij} = \frac{Q_{pi}\hat{Q}'_{aj}F(T)}{\sum_j \hat{Q}'_{aj}F(T)} \tag{3.15}$$

This process of comparing model and survey attraction totals, revising the attractive values (equation (3.14)) and revising the origin-destination matrix (equation (3.15)) continues until a satisfactory row and column balance is achieved. It turns out, in fact, that this adjustment procedure is equivalent to using a doubly constrained model.

Calibration procedures are best explained with a simple example. Figure 3.3 shows a study area partitioned into two residential suburbs (zones 1 and 2) and three employment centres (zones 3, 4 and 5). The daily pattern of trips for home-to-work journeys is summarised as the survey origin-destination matrix, and the transport characteristics are shown as inter-zonal distance matrix. The survey trip-length frequency distribution is plotted and the mean trip length is 3.4 km. The mathematical function of transport impedance selected for the purposes of explaining the calibration steps is the power function of distance. The calibration of the fully constrained model is shown and the sequence of steps in calibrating the other three models are given.

(1) Fully Constrained Model
The fully constrained gravity model to be calibrated is:

$$\hat{Q}_{ij} = \kappa_i \kappa'_j Q_{pi} Q_{aj}/T_{ij}^\alpha$$

where

$$\kappa_i = \left(\sum_{j=3}^{5} \kappa'_j Q_{aj}/T_{ij}^\alpha\right)^{-1} \text{ and } \kappa'_j = \left(\sum_{i=1}^{2} \kappa_i Q_{pi}/T_{ij}^\alpha\right)^{-1}$$

The algorithm to estimate α and the zonal balancing factors (κ_i and κ'_j) follows.

Step 1: Select any initial value of the calibration parameter and calculate the zonal balancing factors. For example, setting $\alpha = 1$, and setting all κ'_j to unity. Balancing factors are solved by an iterative procedure, by assuming any initial value for either κ_i or for κ'_j. The initial value chosen is immaterial because the balancing factors converge (Evans, 1970). To start the iterative process, assume $\kappa'_j = 1.0$ and calculate κ_i:

$$\kappa_1 = \left\{ \frac{(1.0)550}{3} + \frac{(1.0)200}{2} + \frac{(1.0)250}{5} \right\}^{-1} = 0.003000$$

$$\kappa_2 = \left\{ \frac{(1.0)550}{3} + \frac{(1.0)200}{5} + \frac{(1.0)250}{4} \right\}^{-1} = 0.003499$$

Calculate κ'_j given the above value of κ_i:

$$\kappa'_3 = \left\{ \frac{(0.003)300}{3} + \frac{(0.003499)700}{3} \right\}^{-1} = 0.8958$$

$$\kappa'_4 = \left\{ \frac{(0.003)300}{2} + \frac{(0.003499)700}{5} \right\}^{-1} = 1.0640$$

$$\kappa'_5 = \left\{ \frac{(0.003)300}{5} + \frac{(0.003499)700}{4} \right\}^{-1} = 1.2621$$

This ends the first iteration. The second iteration involves substituting the revised values of κ'_j into the formula to recalculate κ_i:

$$\kappa_1 = \left\{ \frac{(0.8958)550}{3} + \frac{(1.0640)200}{2} + \frac{(1.2621)250}{5} \right\}^{-1} = 0.002996$$

$$\kappa_2 = \left\{ \frac{(0.8958)550}{3} + \frac{(1.0640)200}{5} + \frac{(1.2621)250}{4} \right\}^{-1} = 0.003501$$

and these values of κ_i are used to calculate κ'_j, and so on. The values of κ_i in the third iteration equal the values of κ_i in the second iteration, so convergence is achieved. The balancing factors thus calculated are: $\kappa_1 = 0.002996; \kappa_2 = 0.003501; \kappa'_3 = 0.8957; \kappa'_4 = 1.0644; \kappa'_5 = 1.2619$.

Step 2: Substitute the balancing factors into the fully constrained equation given above and calculate the model origin-destination matrix.

Step 3: Calculate the mean trip length of all trips in the model origin-destination matrix. In the example, the mean trip length is 3.4 km.

Step 4: Compare the model and survey mean trip lengths and adjust the calibration parameter (if necessary) *upwards* if the model mean trip length is too *large*, and *downwards* if the model mean trip length is too *small*.

Step 5: Select another calibration parameter and repeat steps 1 to 4 until the calibration criterion is satisfied. In this example, no parameter adjustment is needed and so the estimated inter-zonal pattern of work journeys is:

	3	4	5	$\hat{Q}_{pi}$
1	147.6	95.7	56.7	300.0
2	402.4	104.3	193.3	700.0
$\hat{Q}_{aj}$	550.0	200.0	250.0	1,000.0

(2) Unconstrained Model

The unconstrained gravity model to be calibrated is:

$$\hat{Q}_{ij} = \kappa Q_{pi} Q_{aj} / T_{ij}^{\alpha}$$

where the objective is to estimate the two parameters α and κ.

Step 1: Select any initial value of the calibration parameter α and calculate: $Q_{pi}Q_{aj}/T_{ij}^{\alpha}$ for all $\hat{Q}_{ij}$.

Step 2: Draw an unscaled model origin-destination matrix and calculate total number of inter-zonal trips.

Step 3: Calculate κ by dividing the total number of inter-zonal trips in the survey origin-destination matrix by the total number of inter-zonal trips in the model origin-destination matrix.

Step 4: Multiply the unscaled model origin-destination matrix by κ to obtain the model matrix, given the assumed value of α.

Step 5. Calculate the mean trip length of all trips in the model origin-destination matrix.

Step 6: Compare the model and survey mean trip lengths and adjust the calibration parameter (if necessary) *upwards* if the model mean trip length is too *large*, and *downwards* if the model mean trip length is too *small.*

Step 7: Select another calibration parameter and repeat Steps 1-6 until the calibration criterion is satisfied. (In this example, the parameters are $\alpha = 0.97$ and $\kappa = 0.00321$.)

(3) Production-constrained Model

The production-constrained gravity model to be calibrated is:

$$\hat{Q}_{ij} = \frac{Q_{pi}Q_{aj}/T_{ij}^{\alpha}}{\sum\limits_{j=3}^{5} Q_{aj}/T_{ij}^{\alpha}}$$

Step 1: Select any initial value of the calibration parameter and calculate: $\sum\limits_{j=3}^{5} Q_{aj}/T_{ij}^{\alpha}$ for all i origin zones.

Step 2: Calculate the model origin-destination matrix directly from the production-constrained equation given above.

Step 3: Calculate the mean trip length of all trips in the model origin-destination matrix.

Step 4: Compare the model and survey mean trip lengths and adjust the calibration parameter (if necessary) *upwards* if the model mean trip length is too *large*, and *downwards* if the model mean trip length is too *small*.

Step 5: Select another parameter and repeat Steps 1–4 until the calibration criterion is satisfied.

(4) Bureau of Public Roads Model

Step 1: Define the transport impedance intervals for the friction factors. Here, suitable intervals are distance $T = 2, 3, 4$ and 5.

Step 2: Assume inital values for $F(T)$, for all values of T, and calculate the model origin-destination matrix.

Step 3: Compare the shape of the trip-length frequency distribution with the survey trip-length frequency distribution. This is done by comparing the number of trips in each transport impedance interval.

Step 4: Revise the initial set of friction factors using equation (3.13), recalculate the model origin-destination matrix, and compare model and survey trip-length frequency distributions.

Step 5: Return to step 4 and continue until the calibration criterion is satisfied. In this example, the friction factors are: $F'(2) = 1.755$; $F'(3) = 1.0; F'(4) = 1.116$ and; $F'(5) = 0.698$.

3.3.3 Model Validation

An important question to ask is how accurate are the estimates of inter-zonal traffic when compared with the survey pattern of inter-zonal traffic, and how close is the shape of the model trip-length frequency histogram to the survey histogram (Black and Salter, 1975b). A number of standard statistical tests are helpful.

(1) A widely used measure is the correlation coefficient (equation (3.6)), where the two variables are Q_{ij} and $\hat{Q}_{ij}$, for all $i - j$ pairs. However, the correlation coefficient is insensitive over a wide range of calibration parameters, as noted by Wilson *et al.*, (1969), and Black and Salter (1975b), and 'erroneous', since it measures the degree of linear dependence between two random variables, and $\hat{Q}_{ij}$ is a non-random variable (Wilson, S., 1976, p. 344).

(2) The chi-square test for model and survey origin-destination matrices is:

$$\chi^2 = \underset{i\ j}{\Sigma\Sigma}(Q_{ij} - \hat{Q}_{ij})^2 / Q_{ij} \qquad (3.16)$$

or for trip-length frequency distributions with transport intervals, T:

$$\chi^2 = \underset{T}{\Sigma}\{Q(T) - \hat{Q}(T)\}^2 / Q(T) \qquad (3.17)$$

The value of χ^2 is computed and compared with some critical value, which is obtained from a standard table of the chi-square distribution. The number of 'degrees of freedom' in this table are the total number of $i - j$ pairs minus one, or the number of intervals, T, minus one. If χ^2 is less than the critical value, at a specified level of statistical significance, the interpretation is a model with a 'good fit'. When using equation (3.16) it may be necessary to aggregate zones so as to avoid small numerical values of inter-zonal traffic—the degrees of freedom should be reduced accordingly. Pitfield (1978) describes an algorithm for the chi-square test on trip matrices.

(3) The Kolmogorov-Smirnov test is a standard non-parametric method (Hoel, 1962, pp. 345-9) suitable for solving goodness-of-fit problems between the cumulative survey and model trip-length distributions.

(4) A measure of the plausibility of the model having generated the survey data is the relative likelihood (Wilson, S., 1976, p. 345):

$$RL = \prod_{i,j=1}^{n} (p_{ij}/\hat{p}_{ij})^{Q_{ij}} \qquad (3.18)$$

where

p_{ij} = the survey probability of a trip from zone i to zone j, calculated from $p_{ij} = Q_{ij}/\underset{i\ j}{\Sigma\Sigma}Q_{ij}$;

$\hat{p}_{ij}$ = the model probability of a trip from zone i to zone j, calculated from $\hat{p}_{ij} = \hat{Q}_{ij} / \sum_i \sum_j \hat{Q}_{ij}$;

RL = the relative likelihood measure; and

Π is the symbol for multiplying a series of particular values for the specified algebraic expression.

High values of RL indicate a good model, and the highest value means that the data arose from that model with more plausibility than any other.

3.4 Modal Split

Mode-choice modelling attempts to estimate the total amount of patronage on the different transport modes and to indicate the spatial pattern of this demand. Modal split is the allocation of the total survey traffic to the separate modes, but is often a term used in the more limited sense to mean the bi-partition of traffic into public and private transport modes. The information available for building mode-choice models is the observed modal split, characteristics of the travelling population and operational characteristics of the competing urban transport modes.

3.4.1 Four Types of Mode-choice Model

A rudimentary explanation of transport mode selection contained in Chapter 1 indicated that travellers prefer the least unattractive alternative. Decisions are made by comparing the operational characteristics of alternative urban transport modes, but this bland statement now warrants amplification. There are subjective factors, such as reliability (Golob *et al.*, 1972), convenience (Spear, 1976), comfort and safety (Hartgen and Tanner, 1971), which contribute to any assessment of the attractiveness or otherwise of each transport mode. Furthermore, socio-economic characteristics of the traveller, his attitudes (Golob *et al.*, 1979) and the type of journey all play a part in undermining any simplified explanation of mode choice.

The complexity of factors make it difficult to model mode choice realistically (Schocken, 1968) and consequently a variety of modal-split models have been developed. They can be classified according to their *position* in the modelling sequence (Williams, H., 1977). Figure 3.4 indicates that modal split can be positioned in one of four places: (a) combined with traffic generation; (b) between traffic generation

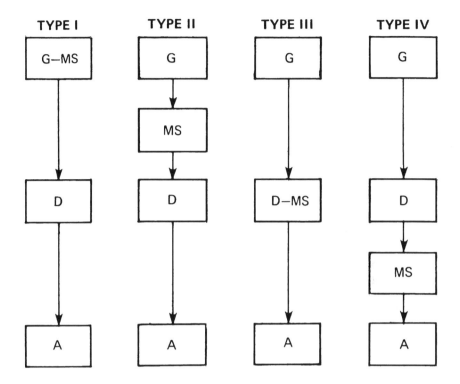

TYPE I   **TYPE II**   **TYPE III**   **TYPE IV**

G — Trip—Generation
D — Trip—Distribution
MS—Modal—Split
A — Traffic Assignment

Figure 3.4: Alternative Positions for Modal-split Analysis

and traffic distribution ('trip-end' models); (c) combined with traffic distribution ('gravity type' models); and (d) between traffic distribution and traffic assignment ('trip interchange' models). For convenience these are called Type I, Type II, Type III and Type IV modal-split models respectively.

Type I modal-split models are formulated as special cases of either the linear regression model or the category analysis model of traffic generation. The dependent variable in the multiple linear regression model is specified as the amount of traffic produced by an origin zone by each transport mode:

$$\hat{Q}_{pi(m)} = \alpha + \beta_1 L_{o1i} + \beta_2 L_{o2i} + \ldots \beta_z L_{ozi} \qquad (3.19)$$

or as the amount of traffic attracted to a destination zone arriving on
each transport mode:

$$\hat{Q}_{aj(m)} = \alpha + \beta_1 L_{d1j} + \beta_2 L_{d2j} + \ldots \beta_z L_{dzj} \qquad (3.20)$$

where

$\hat{Q}_{pi(m)}$ = estimate of the amount of traffic produced by zone i per unit time by transport mode m;

$\hat{Q}_{aj(m)}$ = estimate of the amount of traffic attracted to zone j per unit time by transport mode m;

$L_{o1i}, L_{o2i}, \ldots L_{ozi}, L_{d1j}, L_{d2j}, \ldots L_{dzj}$

= explanatory (independent) land-use variables for origin zone i and destination zone j respectively;

α = regression constant; and

$\beta_1 \ldots \beta_z$ = partial regression coefficients.

The explanatory variables and the parameters will be different (a) for
productions and attractions, and (b) for each transport mode, and if
we were being more explicit in notation, an (m) could be added on the
right-hand side, but here we drop it for simplicity. In the case of two
modes, the equation would usually be estimated for one mode, and
the other obtained by subtraction from the total. In the case of more
than one, the 'last' mode would be obtained in this way.

For category analysis, households are cross-classified by car ownership, size and income in the usual way, but the mean traffic production
by each transport mode is calculated for each cross-classification from
the home interview survey. Zonal traffic production for each mode is
estimated by (a) multiplying the mode-specific trip rate for each cross-
classification by the number of households in each cross-classification
that are located in the zone, and (b) summing the mode-specific traffic
for all household categories in the zone.

Type II modal-split models were among the first to be developed
in the USA because the transport studies of the 1950s and 1960s
were concerned primarily with highway planning, and 'the earlier in
the process that transit trips could be estimated and removed from
further consideration, the more efficient would be the resulting highway
travel forecasting process' (Stopher and Meyburg, 1975, p. 177). There
are numerous examples of Type II models in the literature (Fertal
et al., 1970; Stopher and Meyburg, 1975, pp. 176–85), but the principle
is straightforward: graphical diversion curves. The percentage of
zonal traffic using one transport mode, say public transport, is plotted
against a selected land-use variable, such as zonal car ownership. A

best-fit line or curve is drawn through the data points, and this represents the empirical diversion curve.

Type III modal-split models exploit the gravity model with an exponential function of transport impedance to estimate the amount of inter-zonal traffic on each transport mode (Wilson, 1969). The modal-split formulation, derived from entropy-maximising methods (Wilson, 1970, Chapter 2), in the case of two transport modes ($m = 1$, and $m = 2$) is:

$$\hat{Q}_{ij(1)} / \sum_{m=1}^{2} \hat{Q}_{ij(m)} = \frac{1}{1 + \exp\{-\beta(T_{ij(2)} - T_{ij(1)})\}} \qquad (3.21)$$

where

$\hat{Q}_{ij(1)}$ = estimate of the number of trips from zone i to zone j by transport mode 1;

$\hat{Q}_{ij(m)}$ = estimate of the number of trips from zone i to zone j by transport mode m;

$T_{ij(1)}$ = transport impedance from zone i to zone j by transport mode 1;

$T_{ij(2)}$ = transport impedance from zone i to zone j by transport mode 2; and

β = gravity model calibration parameter.

For example, a fully constrained gravity model has been calibrated with a parameter $\beta = 0.05$ for a hypothetical study area. There are two transport modes linking zone 1 and zone 2, zone 3 and zone 4: private transport ($m = 1$) and public transport ($m = 2$). The travel times from a survey are:

zone-to-zone	1 – 2	1 – 3	1 – 4
private transport (minutes)	20	25	25
public transport (minutes)	10	30	60

The proportions of inter-zonal travel by private transport are: from zone 1 to zone 2, 0.38; from zone 1 to 3, 0.56 and; from zone 1 to 4, 0.85.

Type IV modal-split models take the form of diversion curves, multiple linear regression equations, or a variant of equation (3.21). A comprehensive investigation of the split of traffic between private and public transport as represented by diversion curves was conducted by the Traffic Research Corporation (Hill and Von Cube, 1963). The transport variables included in the analysis were: relative transport costs—the ratio of fares on public transport to the out-of-pocket

motoring expenses (petrol, oil and parking fees); relative travel times—
the ratio of 'door-to-door' travel times by mode; and the 'service ratio'
—the ratio of the total time spent walking, waiting and transferring
on public transport to the time spent parking and walking to and from
a motor car. Traffic was stratified by journey purpose—work and
'other' purposes. Travellers were divided into five income groups.

Figure 3.5 illustrates one set of diversion curves, from a total of
eighty possible curves developed by the Traffic Research Corporation
for Toronto. The graph shows the proportion of journey-to-work
traffic by public transport plotted against the ratio of travel times by
private and public transport. There are four different curves—one for
each 'service ratio'. The data are for medium- to high-income commuters
(economic status 3) who are faced with a transport cost ratio of 0.25.

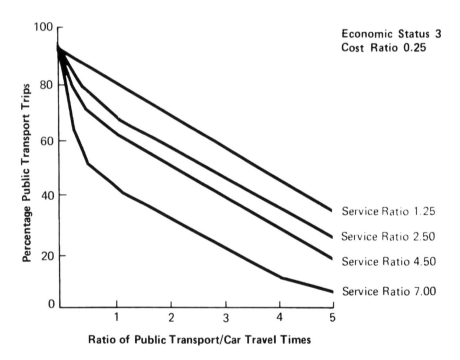

Figure 3.5: Modal-split Diversion Curves, Toronto
Source: Based on Weiner, 1969, Figure 7, p. 15.

The Type IV multiple linear regression model specifies as its
dependent variable the proportion of traffic from zone i to zone j by
one transport mode. The explanatory variables are zonal land-use
variables associated with either the origin or the destination zone,

and transport characteristics between the two zones. The model is:

$$\hat{Q}_{ij(m)}/\sum_m \hat{Q}_{ij(m)} = \alpha + \beta_1 L_{oi} + \beta_2 L_{dj} + \beta_3 T_{ij} \qquad (3.22)$$

where the dependent variable is an estimate of the proportion of total inter-zonal traffic using transport mode m, and where

L_{oi} = land-use variable for origin zone i;

L_{dj} = land-use variable for destination zone j;

α = regression constant; and

$\beta_1 \ldots \beta_3$ = partial regression coefficients.

The model specification may include two (or more) land-use variables both for the origin and destination zones, and two (or more) transport variables.

The variant of the Type III modal-split model arises from 'an observation that the parameter [in equation 3.21] . . . is made to do a lot of work in the model' (Wilson, 1974, p. 143). The revised formulation in the case of two transport modes is:

$$\hat{Q}_{ij(1)}/\sum_{m=1}^{2} \hat{Q}_{ij(m)} = \frac{1}{1 + \exp\{-\gamma(T_{ij(2)} - T_{ij(1)}) + \delta_2\}} \qquad (3.23)$$

where

γ = calibration parameter; and

δ_2 = an optional parameter to reflect a mode-specific handicap penalty.

The role of γ in equation (3.23) is similar to the role of β in equation (3.21), as explained by Williams and Senior (1977, p. 466).

3.4.2 Calibration

The parameters of modal-split models based on linear regression analysis are estimated from data by the method of least squares (section 3.2.2). Diversion curves are usually fitted to the data 'by eye'—the smooth curves are sometimes more imagination than the result of scientific analysis—but there are techniques that eliminate arbitrary curve-fitting procedures (Navin and Schultz, 1970). The parameter of the Type III model is taken from a gravity model calibrated on the spatial distribution of traffic (section 3.3.2).

There are alternative methods that estimate the parameter γ in equation (3.23), although, as noted by Senior and Williams (1977, p. 404), their results may differ by a factor of two. Dorrington (1971) suggests a method that gives acceptable results. The transport impedances for two modes are compared for each inter-zonal linkage and the *difference* in the modal impedances are divided into convenient-sized class intervals $(T = 1, \ldots n)$. The total number of inter-zonal trips and the number of trips by the two transport modes which fall into each of these intervals are drawn as three separate frequency distributions. An initial parameter is assumed for equation (3.23), the modal share of inter-zonal trips calculated, and the results are drawn as frequency distributions. The objective is to obtain a match between the shape of the model and survey frequency distributions: γ is varied until the following measure (s) is minimised:

$$s = \left[\sum_{T=1}^{n} \left\{ \sum_{m=1}^{2} Q_{T(m)} \left(\frac{Q_{T(m)}}{\sum\limits_m Q_{T(m)}} - \frac{\hat{Q}_{T(m)}}{\sum\limits_m Q_{T(m)}} \right) \right\}^2 / \sum_m \sum_T Q_{T(m)} \right] \quad (3.24)$$

where

$Q_{T(m)}$ = number of observed trips by mode m in transport impedance (difference) interval T; and

$\hat{Q}_{T(m)}$ = model estimate of the number of trips by mode m in
transport impedance (difference) interval T.

The 'short' notation implies that the summation is over two transport modes ($m = 1$ and $m = 2$) or over all T transport-impedance class intervals.

3.4.3 Model Validation

Goodness-of-fit methods discussed previously are also applicable to the verification of modal-split models. Model validation is rarely reported, not least because of the inaccuracies which might expose the limitations of models of mode choice. When verification is mentioned, the goodness-of-fit is sometimes poor, as illustrated by four studies.

Estimates of mode-specific inter-zonal traffic in Toronto had a mean error of 46 per cent, and for some individual zone pairs the difference between the observed values and the model estimates ranged from plus 1,000 per cent to minus 83 per cent (Sosslau *et al.*, 1964, Table 5, p. 13). In an analysis of journey-to-work mode choice in Coventry, diversion curves gave estimates such that 'individual origin-destination erros would be incurred of a magnitude that would not be acceptable' (Wilson, F., 1967, p. 92) and multiple linear regression models were equally unacceptable (ibid., p. 84). A Type IV multiple linear regression equation, with a multiple correlation coefficient of 0.87, estimated a mode share on public transport of more than 100 per cent for travel from the Bronx to Manhattan, and within Manhattan (Zupan, 1968, pp. 15-16). A study of journey-to-work travel in Leeds found that the Type III model was 'highly unsatisfactory', although a post-distribution model (equation (3.23)) fitted the data fairly well (Senior and Williams, 1977, Figure 2, p. 405).

3.5 Traffic Assignment

The purpose of traffic assignment modelling is to replicate the amount of traffic on the routes of an urban transport network. Questions are rarely asked in surveys of actual route selection, so the information used to check the reliability of traffic assignment models is roadside traffic counts or counts of public transport patronage on different routes. As there are rarely alternative routes for the user of public transport, the following models apply to the route choice of motorists or lorry drivers.

3.5.1 Three Models of Route Choice

To introduce traffic assignment models, consider the unlikely situation where a journey is made, but there are no other people travelling at that time—this unrealistic assumption is relaxed later in the discussion of capacity restraint. Because there are no other travellers competing for transport, the transport impedances on the network are equivalent to 'zero flow' travel times, or generalised costs.

The modelling of route choice is a two-part process: (a) the identification of the behavioural rationale behind route choice; and (b) the analytical capability to represent the route-choice mechanism.

Three hypotheses concerning traveller behaviour have been proposed:

(1) Travellers have perfect information and act rationally by choosing the route which minimises distance, travel time or generalised cost (Martin, Memmott and Bone, 1961). This is called the *all-or-nothing* assignment because all traffic from an origin zone to a destination zone follows the minimum path through the transport network. No traffic takes any other route.

(2) Travellers have imperfect information about the alternative feasible routes; each traveller 'perceives' travel times, or costs, in a slightly different way. Although each traveller chooses what is considered to be the best route, different perceptions about the minimum path lead to the multi-path assignment (Burrell, 1969).

(3) Travellers assess factors other than minimising distance, travel time or cost, such as the most familiar route, a scenic route or a safe route (Tagliacozzo and Pirzio, 1973). This recognition of individual random preferences leads to the probabilistic assignment (Dial, 1971), where there is a choice probability associated with each alternative route.

Traffic assignment models include both the representation of the transport network and the analytical representation of these hypotheses. The models are best explained with the aid of a simple example. Figure 3.6 shows that part of a transport network which connects two zones labelled *i* and *j*. The network nodes are numbered consecutively from 1 to 5 for identification purposes. The numbers in brackets are the directional ('zero flow') travel times in minutes on each link.

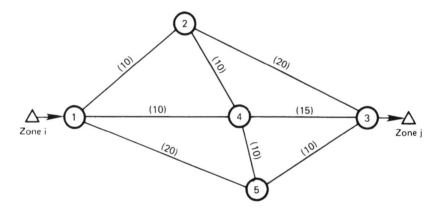

Figure 3.6: Transport Network with Link Travel Times

The *all-or-nothing* assignment specifies that traffic from zone *i* to zone *j* follows the minimum travel time path through the network. From Figure 3.6 the minimum route is found by measuring the total travel times by each alternative route and selecting the path with the lowest value. The minimum travel time route, from node 1 to node 4, and then from node 4 to node 3, has a total travel time of 25 minutes and all traffic is assigned to this route.

For complex networks with many origins and destinations an algorithm calculates efficiently the shortest routes through a network. Creighton (1970, pp. 250-4) explains, with a simple example, the main steps of the minimum-path algorithm, but advances in computer technology have led to better network algorithms (Bonsall, 1976; and Van Vliet, 1976).

The *multi-path* assignment, as described by Burrell (1969), assumes that a traveller does not know the correct link travel times but only 'supposed' link travel times. The model generates the 'supposed' travel times: they are drawn at random from a distribution of travel times which has as its mean value the correct travel time. Figure 3.6 gives the correct link travel times; Table 3.3 gives the distribution of 'supposed' link travel times—the correct time, a travel time 20 per cent less than the correct time, and a travel time 20 per cent more than the correct time.

Table 3.3: Distribution of Link Travel Times for an Example of a Multi-path Assignment

Link Travel Times	Network Node to Node Connections							
(Minutes)	1-2	1-4	1-5	2-3	2-4	4-3	4-5	5-3
Correct	10	10	20	20	10	15	10	10
Minus 20 per cent	8	8	16	16	8	12	8	8
Plus 20 per cent	12	12	24	24	12	18	12	12

In this example, the traveller is confined to one of three routes—through nodes 1-2-3, 1-4-3, or 1-5-3. A die is cast to simulate at random the 'supposed' travel times for each traveller: a throw of one or two simulates the correct times (row 1 in the table); a throw of three or four simulates row 2; and a throw of five or six simulates row 3. Consider the following results:

	Route 1	Route 2	Route 3
node-node	1-2-3	1-4-3	1-5-3
die throw	3,5	6,5	4,2
'supposed' times	8,24	12,18	16,10
route travel time	32	30	26

The minimum path is route 3. For a second traveller the outcomes of the die are 1, 5, 3, 2, 3 and 6 and the best path is route 2 at a total travel time of 23 minutes. By repeating these trials, a multi-path assignment emerges for the three alternative routes.

For large networks and many origin and destination pairs it is impossible to analyse each traveller separately as described above, all travellers from one zone have the same 'supposed' link travel times for the assignment.

The *probabilistic* assignment devised by Dial (1971) uses a clever mechanism to obtain the probabilities of route choice: as routes get longer they become less likely to be used and this is reflected with a decreasing probability. Florian and Fox (1976, p. 339) show that the probability that a traveller uses route k, from a feasible subset of routes linking two zones, is:

$$p(k) = \frac{\exp(-\theta T_k)}{\sum_k \exp(-\theta T_k)} \tag{3.25}$$

where

$p(k)$ = probability of choosing route k;

T_k = travel time on route k; and

θ = traffic diversion parameter.

The probabilities associated with each of the feasible alternative routes sum to unity. Calibration of the parameter θ is explained in the next section, but in the following example θ is assumed to be 0.2. Three feasible routes are considered in Figure 3.6: through nodes 1-2-3 (route 1), 1-4-3 (route 2) and 1-5-3 (route 3). Substituting the correct route travel times and parameter value into equation (3.25): $p(1) = 0.212$; $p(2) = 0.576$; and $p(3) = 0.212$. If the total amount of traffic was 1,000 vehicles from zone i to zone j, 212 take route 1, 576 take route 2 and 212 take route 3.

Traffic assignment models containing 'zero-flow' network travel times are unrealistic for most urban planning applications because they ignore the fundamental interaction between traffic and transport (Chapter 1). The three traffic assignment models—*all-or-nothing, multipath* and *probabilistic*—can be reformulated with a capacity restraint algorithm. Each link in the network has associated with it a traffic flow-dependent travel time function, (or a speed-flow function).

The main steps of a capacity restraint algorithm for use with the *all-or-nothing* assignment, for example, are:

(1) represent transport as a network in the usual way, starting initially with 'zero-flow' link travel times;

(2) calculate the minimum travel time paths through the network for each origin zone in turn;

(3) load the origin-destination pattern of traffic on to the network by an *all-or-nothing* assignment;

(4) calculate the assigned traffic on each link;

(5) use the assigned link traffic flows in the equation for flow-

dependent travel times to calculate revised link travel times;

(6) load the original origin-destination pattern of traffic on to the network by the *all-or-nothing* assignment but with the revised network travel times obtained from Step (5);

(7) return to Step (4) and continue until the assigned traffic and the resultant travel times stabilise.

It is common practice at Steps (3) and (6) to load only a proportion of the origin-destination pattern of traffic, revise the network travel times and then load another portion of traffic. Each subsequent incremental loading involves different minimum paths for the *all-or-nothing* assignment. The merits of incremental against total loading (as described in the above algorithm) have been debated. The incremental method for traffic assignment has 'undesirable properties and in general does not produce equilibrium flows' (Ferland *et al.*, 1975, p. 239). However, according to Weaver and Riley (1971), partial and total loading methods produce similar assignments, but from a computer programming point of view the latter method is more desirable.

The stabilisation of traffic and travel times in the capacity restraint algorithm is of considerable practical importance because it represents convergence towards an equilibrium solution. One criterion of convergence is when link travel times do not change by a specified small percentage from one iteration to the next. Algorithms which converge towards an equilibrium solution have been developed more recently (Daganzo, 1977a).

3.5.2 Calibration

The aim of calibrating *multi-path* or *probabilistic* traffic assignment models is to ensure that the estimated link vehicular flows correspond with roadside vehicle counts. The *multi-path* method requires the specification of the distribution of 'supposed' link travel times, and the *probabilistic* method requires estimation of the traffic diversion parameter θ in equation (3.25).

One calibration criterion suitable for both methods is to minimise the differences between the model link flows and the survey link flows (Robillard, 1975). For a transport network with a total number (n) of individual links (ℓ) the measure (z) to minimise is:

$$z = \sum_{\ell=1}^{n} (Q_\ell - \hat{Q}_\ell)^2 \qquad (3.26)$$

The shape of the distribution of 'supposed' travel times or the parameter θ is adjusted until the above measure is minimised. There are other suitable calibration criteria, such as the maximum likelihood methods (Daganzo, 1977b). A further refinement is to allow θ to take on two or three different values to represent motorists' differing perceptions of route travel times (Daganzo and Sheffi, 1977).

An example of calibrating the *probabilistic* traffic assignment model is given using the calibration criterion in equation (3.26). Referring back to Figure 3.6, roadside traffic counts establish that 215 vehicles per hour use route 1 (nodes 1–2–3), 575 vehicles per hour use route 2 (nodes 1–4–3) and 210 vehicles per hour use route 3 (nodes 1–5–3). Assume an initial value of θ, say $\theta = 0$, and calculate the route probabilities, which are all equal to 0.333. A comparison of model and survey link traffic is:

Network link	1–2	1–4	1–5	2–3	4–3	5–3
Survey traffic	215	575	210	215	575	210
Model ($\theta = 0$)	333	333	333	333	333	333

Solving equation (3.26) the measure of z is 175,234. A search is made for the value of θ that reduces z. The best value (to two decimal places) is $\theta = 0.20$, with $z = 28$, which gives the following assignment:

Network link	1–2	1–4	1–5	2–3	4–3	5–3
Survey traffic	215	575	210	215	575	210
Model ($\theta = 0.20$)	212	576	212	212	576	212

A lower parameter of $\theta = 0.19$ increases z to 388, and a higher parameter of $\theta = 0.21$ increases z to 532.

3.5.3 Model Validation

The rigorous method to check the validity of traffic assignment models is to compare network link assignments with survey counts, and to calculate the percentage difference. Usually this comparison is made only at a screenline, which is an imaginary line drawn across the study area not passing through the city centre (Bruton, 1975, pp. 77–8).

Equilibrium assignments usually produce more satisfactory assignments but at higher computational costs. The disadvantage of the *all-or-nothing* method is that the traffic flow patterns alter considerably with small changes to the link impedances (Mason, 1972, p. 280). Whereas the *all-or-nothing* predicted correctly about half of the routes, the *multi-path* assignment increased this to nearly 90 per cent (Ratcliffe, 1972, p. 529). An equilibrium assignment was accurate for a network of 2,789 links in Winnipeg (Manitoba), but gave poor results on links with fewer than 300 vehicles per hour, and was sensitive to the way the road pattern was represented as a network configuration (Florian and Nguyen, 1976).

3.6 Behavioural Modelling of Traffic

An alternative approach, which emerged during the 1970s and is still being refined (Hensher and Stopher, 1979), is the analysis of individual travel behaviour. A behavioural model 'represents the decisions that consumers make when confronted with alternative choices' (Domencich and McFadden, 1975, p. 4). Travel-related choices include whether to make a journey or not, time of day of travel, destination, transport mode and route.

An important distinction is that behavioural models represent individual travel, whereas the models discussed previously represent 'bundles of individual trips' aggregated into zones. Several advantages follow from this micro-scale analysis (Hensher, 1977c, pp. 81-2). Conceptually, the analysis is more realistic because behavioural patterns of individuals are investigated instead of statistically derived zonal correlations. Behavioural models are calibrated on individual data, thereby making a more efficient use of the survey. Behavioural models avoid the 'ecological fallacies of inference' from grouped data (Kassoff and Deutschman, 1969).

3.6.1 Behavioural Models

A behavioural model is based on a representation of individual choice when faced with alternatives. Common experience suggests that in choice situations a person weighs up the advantages and disadvantages of one course of action against the advantages and disadvantages of the alternatives. The comparison is made on an assessment of the *attributes* of each alternative, such as price or quality. A logical decision is to select the alternative which gives the 'greatest enjoyment', 'the most satisfaction', 'the highest utility', or whatever phrase best describes the derived emotion.

Expressing the same thing in notational terms, assume there are two alternatives a and b, which have a number (n) of relevant attributes (X). The attributes for each alternative are specified in the following two vectors:

$$[X_{1(a)}, X_{2(a)}, X_{3(a)} \ldots X_{n(a)}]$$

$$[X_{1(b)}, X_{2(b)}, X_{3(b)} \ldots X_{n(b)}]$$

An individual's revealed preference function is a function of the derived satisfaction from alternative a, and is stated theoretically as:

$$U_i = f_i(X_{1(a)}, X_{2(a)}, X_{3(a)} \ldots X_{n(a)}) \qquad (3.27)$$

where

$U_i =$ revealed preference function for individual i; and
$f_i =$ a function specific to individual i.

A similar function may be stated for alternative b. Alternative a is chosen if.

$$f_i(X_{1(a)}, X_{2(a)} \ldots X_{n(a)}) > f_i(X_{1(b)}, X_{2(b)} \ldots X_{n(b)})$$

Extending this argument to a multi-choice situation, the alternative with the highest preference function is selected.

The principle of individual choice is extended to formulate a generalised model for the population as a whole. Within the framework of economic rationality and utility maximisation, tastes vary over the population. This is handled in the model by assuming that each individual preference function has two parts: the first is common to a definable subgroup within the population (for example, a socioeconomic group); the second is unique to each individual. Because this second part cannot be observed, the analyst can attach only a probability to any individual choice.

There are alternative functional forms for the choice probabilities for two alternatives (Domencich and McFadden, 1975, pp. 53-65), but a suitable expression, selected partly because of computational convenience, is the binary logit probability model:

$$P_{i(a)} = \frac{\exp\{f(X_{1(a)} \ldots X_{n(a)})\}}{\exp\{f(X_{1(a)} \ldots X_{n(a)})\} + \exp\{f(X_{1(b)} \ldots X_{n(b)})\}} \qquad (3.28)$$

where

$P_{i(a)}$ = probability of individual choosing alternative a;
$f(X_{1(a)} \ldots X_{n(a)})$ = preference function for alternative a;
$f(X_{1(b)} \ldots X_{n(b)})$ = preference function for alternative b; and
$X_1, \ldots X_n$ = relevant attributes.

The probability of choosing the alternative b is equal to $1 - P_{i(a)}$. Equation (3.28) is derived mathematically, as shown by Stopher and Meyburg (1975, pp. 279–81), from the *assumption* that the distribution of the random, second part of the individual preference function is a Weibull distribution. The Weibull frequency distribution, which has the same general bell-shape as a normal distribution except that it is skewed (Domencich and McFadden, 1975, Figure 4.5, p. 62), is chosen because it has some convenient mathematical properties.

One of these properties is to reduce the mathematical complexity of deriving a formulae, equivalent to equation (3.28), for the case of choice amongst more than two alternatives. The probability of an individual choosing alternative a from a set, N, which contain a up to n alternatives, is the multinomial logit probability model:

$$P_{i(a)} = \frac{\exp\{f(X_{1(a)} \ldots X_{n(a)})\}}{\displaystyle\sum_{N=a}^{N=n} \exp\{f(X_{1(N)} \ldots X_{n(N)})\}} \tag{3.29}$$

where

$f(X_{1(N)} \ldots X_{n(N)})$ = preference function for the general case of alternative N.

The application of the above theory of individual choice behaviour to the analysis of urban travel demand is relevant because each traveller is confronted with choices, including to make a journey or not, the time of day, the destination, the transport mode, and the route. Theoretically, choice of land-use activities and transport are made simultaneously from all possible activity-transport combinations, but this is impossible to model satisfactorily. Instead, the choices are separated (Brand, 1974) and sequential models are constructed.

Formulating a model of behavioural travel demand involves the identification of the relevant attributes which influence the individual's decision-making process. Identification and measurement of these attributes leads to the quantification of the preference function. The attributes influencing individual travel behaviour are straightforward to identify because the variables used in aggregate models are relevant also at the individual scale of analysis. The individual's preference function is comprised of the attributes of land-use activity-transport characteristics and individual socio-economic characteristics, and equation (3.27) becomes:

$$U_i = f_i(\mathbf{L}_i, \mathbf{T}_i, \mathbf{S}_i) \tag{3.30}$$

where

U_i = revealed preference function for individual i for land-use activity–transport alternative;

L_i = land-use activity attributes confronting individual i;

T_i = transport attributes confronting individual i; and

S_i = socio-economic characteristics of individual i.

The use of bold variables indicates that each of L_i, T_i and S_i may be a list of variables.

The form of the preference function is assumed to be a linear combination of the relevant attributes of each alternative. Thus, if there is one variable in each group, the equation is

$$f_i(L_i, T_i, S_i) = \alpha + \beta_1 L_i + \beta_2 T_i + \beta_3 S_i \qquad (3.31a)$$

where

$\alpha, \beta_1 \ldots \beta_3$ = calibration parameters.

More generally, if there is more than one variable, a vector notation is appropriate:

$$f_i(\mathbf{L}_i, \mathbf{T}_i, \mathbf{S}_i) = \alpha + \boldsymbol{\beta}_1 \cdot \mathbf{L}_i + \boldsymbol{\beta}_2 \cdot \mathbf{T}_i + \boldsymbol{\beta}_3 \cdot \mathbf{S}_i \qquad (3.31b)$$

The right-hand side of equation (3.31) (a or b as appropriate) is substituted into either the binary logit probability model (3.28) or the multinomial logit probability model (3.29). Calibration is described in the next section, and involves both the identification of the exact variables that enter the preference function and the estimation of the unknown parameters.

An example of the results obtained by calibrating a binary logit probability model of transport mode choice for work travel based on commuters in Sydney who have a choice between car and train is in Hensher (1972). The relevant attributes of the two transport alternatives were found to be door-to-door travel time, waiting time, walking time and transport cost. Because this is a binary choice the variables in equation (3.31) can be expressed as differences. The preference function calibrated is a version of (3.31b) with three transport variables and no land-use or socio-economic variables. For clarity, the individual subscript i associated with the variables has been omitted from the right-hand side of the equation.

$$f_i(T_a - T_b) = 0.254 - 0.010(I_a - I_b) - 0.028(W_a - W_b)$$

$$- 0.015(R_a - R_b) - 0.027(C_a - C_b) \qquad (3.32)$$

where

$f_i(T_a - T_b) =$ preference function for individual i;

$I_a - I_b$ = in-vehicle travel time by car minus in-vehicle travel time by train in minutes;

$W_a - W_b$ = waiting time by car (if any) minus the total waiting time by train in minutes;

$R_a - R_b$ = walking time involved in using the car minus the walking time and transfer time associated with the train, in minutes; and

$C_a - C_b$ = door-to-door cost by car minus the door-to-door travel cost by train in cents.

The negative coefficients indicate that when travel times become longer, or costs become higher by train (the differences between car and train will be negative and increasing) the preference function for car choice has an increasing positive value.

The above expression can be substituted into the binary logit probability model (equation (3.28)), but because differences in transport attributes are specified, equation (3.28) is now written as:

$$P_{i(a)} = \frac{\exp\{f_i(T_{(a)} - T_{(b)})\}}{1 + \exp\{f_i(T_{(a)} - T_{(b)})\}} \tag{3.33}$$

where

$P_{i(a)}$ = probability of individual i taking transport alternative a (motor-car) for the journey to work.

Consider, for example, a group of commuters faced with the following transport attributes in their journey from home to work: in-vehicle time is 10 minutes longer by car; waiting time is 5 minutes more by train; walking and transfer time is 12 minutes longer by train; and the cost is 5 cents more by car. The model estimates that the probability that an individual will choose a, the car, is 0.58.

More complete urban travel demand models incorporating mode choice, time-of-day choice, destination choice and frequency of travel have been developed (Domencich and McFadden, 1975, pp. 165–78). Further reading about such developments is found in Williams (H., 1977).

In summary, the properties of behavioural models of travel demand are:

(1) the probability of choosing one alternative over alternative courses of action is a function of the attributes of the alternatives;

(2) a linear, additive preference function represents the amount of satisfaction that an individual derives from the attributes associated with choice;

(3) the logit transformation guarantees that the probability of choice is restricted to from zero to unity, irrespective of the numerical magnitude of the preference function;

(4) the probability of choice increases as the numerical measure of the attribute of that alternative gets larger relative to the measure for the alternatives. The probability of choice decreases as the numerical measures of the alternative gets smaller relative to the measure of the alternatives (transport travel time, or costs, represent a disutility to the individual traveller).

3.6.2 Model Calibration

Calibration of behavioural models involves testing alternative explanatory variables (attributes) in the preference function, selecting the best combination that explains the choice behaviour, and estimating the parameter associated with each variable (attribute). The model is calibrated using data describing each individual and not each zone.

Conceptually, the calibration of the preference function is similar to estimating the parameters for a multiple linear regression model. However, regression is unsuitable for statistical reasons (Watson, 1974), and the appropriate way to estimate the unknown parameters is by the maximum likelihood method (de Donnea, 1971, Chapter 4). Most applications of the maximum likelihood method would involve the use of a computer programme. The log likelihood (which is equivalent to the likelihood) of observing a given sample containing individuals $(i = 1 \ldots I)$ is:

$$LL = \sum_{i=1}^{I} \log\{1 + \exp(\alpha + \beta_1 L_i + \beta_2 T_i \ldots)\}$$

$$+ \sum_{i=1}^{I} f_{ai}(\alpha + \beta_1 L_i + \beta_2 T_i \ldots) \qquad (3.34)$$

where

LL = log likelihood function; and

f_{ai} = $\begin{cases} 1 \text{ if } a \text{ is chosen by individual } i; \\ 0 \text{ if } a \text{ is not chosen by individual } i. \end{cases}$

The method of maximum likelihood argues that the calculated probabilities of observing the given sample is highest when the calibration parameters have their true values. The highest, or maximum, value of the log likelihood is found by differentiating equation (3.34) with respect to the unknown parameters, and setting the derivatives equal to zero. The solution of these equations gives an estimate of each parameter. For the multiple choice model, maximisation is achieved by a standard iterative process, such as the Newton-Raphson method (Domencich and McFadden, 1975, pp. 121-2).

Although the calibrated model estimates an individual choice probability, the overall probability of choice for any defined group can be easily obtained by summing the individual estimates and dividing the figure by the number of individuals in the group. Aggregation to a spatial, zoning system is less straightforward when the individuals are impossible to identify, and when variables are represented only by zonal means. McFadden and Reid (1974, p. 28) give a formula for calculating zonal probabilities, given a calibrated, disaggregate model, the zonal means for the explanatory variables and the intra-zonal variances of these variables.

3.6.3 Model Validation

Many of the points made in section 3.2.3 are of relevance, but additionally there are three useful goodness-of-fit statistics associated with maximum likelihood methods:

(1) the t-test, which allows statements to be made about the confidence that may be placed in the estimates of the parameters (Hutchinson, 1974, pp. 39-40);
(2) an index analogous to the coefficient of multiple determination, and sometimes called 'pseudo-r^2' which is calculated from the ratio of the value of the log likelihood function, evaluated for the estimated parameters, and the value of the log likelihood function, evaluated for all parameters set to zero (Domencich and McFadden, 1975, p. 123). The index is from zero to one, with the latter indicating a good fit to the data.
(3) the weighted proportion of the number of successful estimates of choice by the model when compared with actual choices made by individuals in the sample.

3.7 Summary

In this chapter, the component parts of what might be collectively termed an 'urban travel demand model' have been isolated. The aim has been to introduce the theoretical structure of the different sub-models, to explain calibration procedures, and to suggest ways that assess how well the results of each model compare with survey data.

Accurate representation of urban travel demand by a systems model is complicated, not least because the system of interest concerns human behaviour. In order to simplify analysis, travel decisions are divided into sequential sub-models of traffic generation, traffic distribution, modal choice and route choice (traffic assignment). Furthermore, analysis of the data may be based on traffic aggregated by zones or on individual trips, as in the case of behavioural models.

Irrespective of the sub-models chosen to represent travel demand, every effort should be made to formulate a model which gives an equilibrium state of the system. The travel times (or costs) resulting from the assignment of travel demand on to the transport network are required to adjust the spatial pattern of traffic, and modal split, thereby ensuring internal consistency between the sub-models. This feedback or cyclical procedure is rarely conducted in practice because of the computational costs, but there are exceptions (section 7.2).

Although systems modelling gives theoretical insights into factors influencing travel demand, the calibration of models is undertaken primarily to provide a sound basis for making traffic forecasts, as explained in the next chapter. The applications to planning practice of aggregate and disaggregate travel demand models are described in Part Two.

4 FORECASTS, PLANS AND EVALUATION

Urban transport plans are formulated in response to anticipated traffic conditions. To consider a wide range of land-use and transport possibilities it is necessary to estimate traffic at various dates in the future and to decide the best way to meet these increases. This chapter explains how traffic forecasts are made, suggests some principles for formulating transport plans, and shows how alternative plans are evaluated. The overall aim of this part of the transport planning process is to advise policy-makers on the alternatives available, the trade-offs involved, the resources required and the likely benefits.

Because traffic is a function of land use, the first three sections provide an elementary introduction to techniques for forecasting population, employment and car ownership at the aggregate urban scale, the zonal scale and the disaggregate household scale. The fourth section explains how travel demand forecasts are calculated from changes in future land-use activity. The assumptions underlying these traffic forecasts and their likely accuracy are discussed. The fifth section suggests some principles to follow when formulating transport plans. Finally, methods used in the economic, social and environmental appraisal of alternative plans are outlined.

4.1 Aggregate Forecasting

Future trends in urban population and employment are the yardsticks to assess the rate of future land development and the space requirements for various land-use categories, whereas future trends in vehicle ownership give an indication of future traffic. The future is uncertain but quantitative techniques provide a sound basis for preparing forecasts in the light of prevailing conditions and assumptions. Alternative methods for making population, employment and car ownership forecasts are explained and their limitations pointed out.

4.1.1 Population

Extrapolation, or trend projection, is straightforward and has the merit of giving sensible estimates without bothering about the underlying factors causing a change in population. For any study area where population records have been kept, a graph is drawn of population each year (or census year). If the population axis is a logarithmic scale the 'best-fit' curve or line through the data points shows the rate of population growth. Extension into the future gives population estimates for any future year. Problems occur when historical growth rates fluctuate over time, because it is difficult to identify the trend that should be projected forwards.

The two ways of establishing the 'best-fit' line or curve are either to sketch it 'by eye' and continue the trend or, more scientifically, to fit a mathematical function of population against time. The function to use depends largely on the historical growth patterns, but the common ones are: the straight line; the quadratic; the exponential; and the logistic. The parameters of the function are estimated by the method of least squares.

Population projections by the cohort-survival method (Benjamin, 1968; Cox, 1970) recognise that the components of change are related to births, deaths and migration. Basically, the present age structure of the population is modified over time by current or anticipated birth rates, death rates and migration rates. The minimum data requirements are: (a) the age structure of the commencing population; (b) a set of age-specific survival factors; and (c) a set of age-specific fertility factors.

The projection of a commencing population is illustrated in Figure 4.1. Here, the 'survival factor' is the probability of being alive in the next year. Thus, the population in year 0 (the start) for each age is multiplied by the appropriate age-specific 'survival factor' to give the population in year 1: this product is entered at one age lower and at one step to the right in Figure 4.1. This 'ageing' process continues for each year of the analysis. Births are introduced for each projection year. The anticipated number of infants aged less than one year in year N is estimated by multiplying the age-specific population in year $N - 1$ by its appropriate age-specific fertility factor. (The fertility factor for people aged fifteen or less is assumed to be zero.) The total number of births in the twelve months for all age groups between year $N - 1$ and N are added to the population, and then 'aged' as described above.

This description refers to age-specific fertility rates, but the principle is easily extended to calculate future births based on female fertility rates. Additional refinements can be made by including marital status and the migration component (Cox, 1970, pp. 257-72). Usually, the population is divided into age groups, or cohorts—0 to 5, 5 to 10 and so on—and projections are made for successive five-year intervals.

AGE	STARTING POPULATION (YEAR 0)	SURVIVAL FACTOR	POPULATION IN YEAR 1	POPULATION IN YEAR 2	POPULATION IN YEAR 3
.					
.					
.					
23	$^0P_{23}$	S_{23}			
24	$^0P_{24}$	S_{24}	$^0P_{23} \cdot S_{23} = {}^1P_{24}$		
25	$^0P_{25}$	S_{25}	$^0P_{24} \cdot S_{24} = {}^1P_{25}$	$^1P_{24} \cdot S_{24} = {}^2P_{25}$	
26	$^0P_{26}$	S_{26}	$^0P_{25} \cdot S_{25} = {}^1P_{26}$	$^1P_{25} \cdot S_{25} = {}^2P_{26}$	$^2P_{25} \cdot S_{25} = {}^3P_{26}$
27			$^0P_{26} \cdot S_{26} = {}^1P_{27}$	$^1P_{26} \cdot S_{26} = {}^2P_{27}$	$^2P_{26} \cdot S_{26} = {}^3P_{27}$
28				$^1P_{27} \cdot S_{27} = {}^2P_{28}$	$^2P_{27} \cdot S_{27} = {}^3P_{28}$
29					$^2P_{28} \cdot S_{28} = {}^3P_{29}$
.					
.					
.					

Figure 4.1: Population Projection Method
Source: Based on Cox, 1970, Figure 15.1, p. 254.

Age-specific mortality, fertility and migration rates are adjusted accordingly for each cohort. Any practical application of the cohort-survival method involves lengthy computations and the tedium of manual calculations is overcome by using a computer programme (Baxter 1976, pp. 90-6).

A proper accounting framework (Rees and Wilson, 1977) is essential if each member of the population, past and future, is to be accounted for in a temporal and spatial way. Briefly, the existing population is in an *initial* state, defined by age, sex and geographical location, and moves into a *different* state at the next time period—through the ageing process or the migration process. The possible states are the rows and columns of a matrix whose elements are the number of people who move from the *initial* state to the newly defined state. Division of each element by its row total is the rate at which population changes state.

Normally, the accounting matrix is divided into four sub-matrices containing:

(a) the existing population who 'survive' the time period;
(b) the existing population who die during the time period;
(c) those born during the time period, and who 'survive'; and
(d) those born during the time period, but who die.

Separate rates are measured for each quadrant of the matrix.

The population forecasting model, based on the principles of accounting, applies either the transition rates calculated from historical data or postulates transition rates. One advantage of the accounting framework is that explicit definitions of all rates—survival, birth and migration—are provided. Consequently, although forecasts may be based on rates crudely derived they are chosen 'in a way that ensures that the right kind of rate (from an accounting point of view) is being measured' (Rees *et al.,* 1977, p. 115). Another advantage is that regional development policies can be included in the model by adjustments in the migration rates.

4.1.2 Employment

Most employment forecasts used for transport planning purposes are extrapolations of historical trends. Usually, the trends are made for separate employment and occupation categories to reflect structural changes in the primary, secondary, tertiary and quaternary sectors of the economy. Alternatively, total employment is derived directly from population forecasts by extrapolating trends in labour-force participation rates.

A better understanding of the growth of the urban economy is achieved if the economic-base method is used to make employment forecasts. Basically, the method applies the theory of international trade to urban areas and fluctuations in the levels of 'urban exports' determine the amount of urban economic activity, including the likely amount of employment. Economic activity is divided into *basic* and *non-basic* components. *Basic* components are 'the export activities of the community that bring in its net earnings and enable it to continue as an independent economic entity' and *non-basic* components are 'completely local and involve no exports beyond the predetermined limits of the area' (Andrews, 1953, pp. 263-4).

The growth hypothesis is that the *basic* sector is the engine of growth and that changes in *non-basic* activities follow in an entirely predictable way. This relationship between *basic* and *non-basic* employment is specified as:

$$\hat{E}_n = \epsilon E_b \qquad (4.1)$$

where

$\hat{E}_n =$ estimate of the amount of *non-basic* employment;

$E_b =$ amount of *basic* employment; and

$\epsilon =$ economic-base multiplier.

The multiplier is calculated from historical data and is the ratio *non-basic* to *basic* employment in the study area. The forecasting method assumes that the future level of *basic* employment can be forecasted exogenously. Application of the multiplier will give forecasts in the *non-basic* sector.

Input-output analysis takes a more disaggregate view of the urban economy than the economic-base method. Essentially, input-output is a method of tracing and using information about all those transactions between buyers and sellers (usually producers and consumers) located within the study area. The input-output transaction matrix is a double-entry accounting framework recording the annual purchases and sales of economic sectors which form the rows and columns of the matrix.

'Technical' or 'marginal input' coefficients are estimated from the data contained in the matrix (Hirsch, 1973, p. 201) and they allow a systematic tracing of the effects of (future) production by one sector on the production of other sectors. By assuming productivity ratios, estimates of employment by sector can be derived. The overriding problem of obtaining urban employment forecasts in this way is the lack of suitable local area data (Smith and Leigh, 1977).

4.1.3 Car Ownership

National or urban-wide forecasts of motor car ownership are extrapolations based on a logistic curve fitted to historical data on the growth of vehicles *per capita* over time. The curve has a characteristic S-shape because the ownership of durable goods, such as motor vehicles, reaches a finite asymptote, or saturation level (Tanner, 1962, p. 264). The algebraic expression for the logistic curve is:

$$C_t = \frac{\sigma}{1 + \alpha \exp(-\beta \sigma t)} \qquad (4.2)$$

where

C_t = car ownership per person at time t;
σ = saturation level to which car ownership becomes asymptotic; and
α, β = parameters.

The parameters in the denominator are estimated from:

$$\alpha = (\sigma - C_0)/C_0 \tag{4.3}$$

$$\text{and } \beta = g_0/(\sigma - C_0) \tag{4.4}$$

where

g_0 = rate of growth of car ownership at time 0; and
C_0 = car ownership per person at time 0.

One method of estimating the saturation level is to plot the percentage change (usually an increase) in cars per person on the vertical axis against cars per person on the horizontal axis for historical time-series data. By the method of least squares a line is fitted to these data points which is extrapolated beyond the range of the data until it crosses the horizontal axis of the graph. The value of cars per person shown is the estimate of saturation because it is where the percentage change in cars per person is zero. Future car ownership is determined by the saturation level, the numerical value of which has been widely debated in the UK (Tanner, 1974; Fowkes and Button, 1977).

More recent work (Tanner, 1977) has led to a method which allows for variations in the approach path towards saturation by incorporating assumptions about motoring costs and incomes:

$$C_t = \frac{\sigma}{1 + \beta(I_t/I_0)^{\gamma_1 \sigma} (M_t/M_0)^{\gamma_2 \sigma} \exp(\gamma_3 \sigma t)} \tag{4.5}$$

where all terms have been defined previously except for

I_t = income per head at time t;
I_0 = income per head at time 0;
M_t = motoring costs at time t;
M_0 = motoring costs at time 0; and
$\gamma_1, \gamma_2, \gamma_3$ = parameters.

A further development is the person-based car ownership forecasting model (Mullen and White, 1977), where the finite asymptote is replaced by a saturation level for each person type:

$$C_s = \frac{\sigma_s}{1 + \exp(-\beta_1 X_1 - \beta_2 X_2 \ldots - \beta_n X_n)} \qquad (4.6)$$

where

C_s	=	average car-ownership per person-type s;
σ_s	=	saturation level of person type s;
$X_1 \ldots X_n$	=	explanatory variables, such as income and public transport quality; and
$\beta_1 \ldots \beta_n$	=	parameters.

Saturation levels varied from 0.5 cars per person for pensioners to 0.9 cars per person for male workers.

4.2 Zonal Forecasting

Traffic forecasts are based not only on projections of population, employment and car ownership but on the zonal distribution of these, and other, land-use activities. The preparation of forecasts at this detailed spatial resolution depends on the traditional skills of the town planning profession (McLoughlin, 1969, p. 237). Statutory planning, through land-use zoning and development controls, is almost the only purposeful way of influencing the way a city grows and develops.

The very notion of planning entails taking decisions in advance, but it is impossible to predict accurately the demand for land for different purposes and in different locations, and therefore land-use proposals in statutory planning schemes are often an uneasy balance between what would be likely to occur in the ordinary course of events and what the planners are able to persuade the planning authority *should* be allowed to happen. Consequently, the planner faces a dilemma between *prescribing* and *forecasting* land use (Black, 1975, pp. 318-19).

In preparing zonal population forecasts, due recognition must be given to the dynamics of residential change—growth, stabilisation and decline or redevelopment. The holding capacity is a suitable method of prescribing population for small geographical areas, such as traffic zones. Suitable land for future residential purposes can be assessed from land-use surveys. The number of dwelling units which might be accommodated is calculated from permissible residential densities, and assumptions about future occupancy rates allow likely estimates of population to be made.

As employment location is less susceptible to planning control than housing, forecasts of its spatial distribution are based on an assessment of land suitable for future industry or commerce, and prevailing, or anticipated, employment densities. Other land-use activities are difficult to predict unless the planner is guided by a statutory plan or standards for neighbourhood design.

An alternative approach is to design a land-use model that predicts future zonal distributions of population and employment. Following the work of Lowry (1964; Batty, 1975) the urban economy is divided into three sectors: *basic* employment, *non-basic* employment and households. The future level and zonal location of *basic* employment is determined by considerations that are exogenous to the model. In outline, the model has an iterative structure that estimates the number of households from *basic* employment and locates them according to accessibility to jobs, and then estimates the amount of *non-basic* (service) employment from the number of households and allocates this to zones according to accessibility considerations. As population and *non-basic* employment builds up with each model iteration, constraints imposed by the holding capacity of each zone keep the zonal estimates within sensible bounds.

Modelling of car ownership is a more widely accepted approach towards zonal forecasting. Regression analysis is one approach: zonal car ownership (or the average number of cars per household) is defined as the dependent variable and the explanatory power of a number of land-use variables are examined (see section 3.2.1). Zonal car ownership is often related statistically to characteristics such as income and residential density. The coefficients fitted in the model are assumed to remain constant over time and car ownership forecasts are obtained by substituting the future zonal values of the explanatory variable (which are forecasted exogenously to the model) into the regression equation.

An alternative approach is to divide households into three categories —non-car-owning, single and multiple car-owning households—and to develop separate forecasting equations for each (Bates *et al.*, 1978). The function for single car-owning households is:

$$C(1\|I) = \frac{\sigma_1}{1 + \exp(-\alpha_1 - \beta_1 \log I)} \tag{4.7}$$

for multiple car-owning households the function is:

$$C(2\|I) = \frac{\sigma_2}{1 + \exp(-\alpha_2 - \beta_2 I)} \tag{4.8}$$

and the residual proportion are zero-car households.

In equations (4.7) and (4.8):

$C(n\|I)$ = proportion of households owning n cars given a house-hold income I;

σ_n = saturation level for households owning n cars;

I = household income; and

α, β = parameters.

The parameters can be estimated by the method of maximum likelihood. A further refinement is to stratify households into six groups according to accessibility to employment. Forecasts of the future proportions of households in these categories require independent forecasts of income and accessibility.

4.3 Disaggregate Forecasting

The distribution of household types *within* zones is required both for category analysis and for any examination of the distributional or equity implications of alternative plans. Zonal traffic production forecasts from the category analysis model require the future number of households in each cross-classification to be predicted for each zone. The 'movement' of a household from a present cross-classification into a new cross-classification is the result of changes to household structure, income or car ownership. Probability distributions represent the shape of the frequency distribution of household incomes, the distribution of car ownership by income group and the frequency distribution of household size (and composition). Alteration to the mean value of these distributions for each zone changes the cross-classifications *within* that zone.

4.3.1 Income

The income distribution is represented by a gamma function:

$$P(I) = \int_a^b \frac{\alpha^{\eta+1}}{\Gamma(\eta + 1)} I^\eta \exp(-\alpha I) dI \qquad (4.9)$$

where

$P(I)$ = probability that a household has income I;

$\displaystyle\int_a^b$ = integral with a as the lower limit of income group and b as the upper limit of income group;

I = household income;
α, η = parameters; and
Γ = gamma function, whose values are obtained from standard tables.

Parameters are estimated from the income distribution observed in the base-year; α is calculated from the ratio of the mean value of the distribution of all households to its variance; and η is calculated from:

$$\eta = \alpha \bar{I} - 1 \qquad (4.10)$$

where
$\bar{I}$ = mean income of households.

Figure 4.2 shows the shape of the income distribution in Toowoomba, Queensland, for 1966, and the gamma function which best fits these data. The parameters of equation (4.9) are $\alpha = 0.873$ and $\eta = 1.62$ (Golding, 1972).

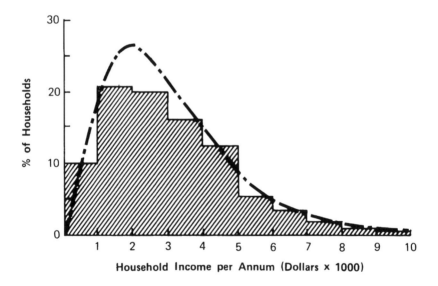

Figure 4.2: Category Analysis Income Distribution
Source: Based on Golding, 1972, Figure 2, p. 311.

For forecasting, a separate forecast is required for future zonal mean household income, often made by assuming a constant growth rate of incomes over time:

$$\bar{I}_t = \bar{I}_0(1 + g)^t \tag{4.11}$$

where

$\bar{I}_t$ = mean household income at time t;
$\bar{I}_0$ = mean household income in survey year 0; and
g = income growth rate per annum.

Future mean income is substituted into a rearranged equation (4.10) to obtain a new value of the parameter α. (The parameter η is assumed to remain unchanged over time). Equation (4.9) with the new parameter is used to calculate this proportion of households with any specified level of income. Sometimes there are practical difficulties in fitting gamma distributions to income data from surveys (Mackinder *et al.*, 1975).

4.3.2 Car Ownership

The car ownership distribution comprises: households with no car, households with one car and multiple car-owning households. As household income increases from zero the proportion of households with no car falls steadily, the proportion of one-car households increases, reaching a maximum in the middle income range, and then declines, and the proportion of multiple car-owning households increases especially in the higher income ranges (Figure 4.3). The distribution representing households with either zero or one car is the conditional probability:

$$\Phi(C\|I) = \alpha_C \, I^{\beta_C} \exp(-\gamma_C I) \tag{4.12}$$

where

$\Phi(C/I)$ = probability of a household owning C cars given an income I;
C = number of cars (either $C = 0$ or $C = 1$); and
$\alpha_C, \beta_C, \gamma_C$ = parameters.

Because the probability must sum to unity, the probability of households owning two or more cars is:

$$\Phi(\geqslant 2\|I) = 1 - \Phi(0\|I) - \Phi(1\|I) \tag{4.13}$$

Income forecasts should be adjusted to account for changes in incomes relative to car prices and equation (4.11) should be replaced by:

$$\bar{I}_t = \bar{I}_0(1 + g - g_c)^t \tag{4.14}$$

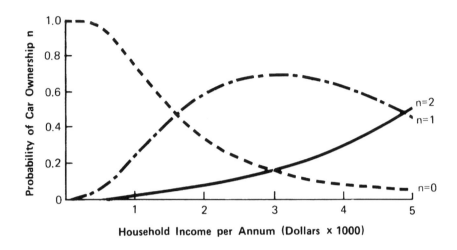

Figure 4.3: Category Analysis Car Ownership Distribution
Source: Based on Golding, 1972, Figure 3, p. 314.

where all terms have been defined previously, except

g_c = growth rate per annum of car prices.

4.3.3 Household Structure

Household structure is constructed from two distributions: the number of persons per household, and the number of employed persons per household. The frequency distribution of the number of persons per household is assumed to follow a Poisson distribution (Figure 4.4):

$$\Psi(N) = \frac{\exp(-\bar{N})\,\bar{N}^{N-1}}{(N-1)!} \qquad (4.15)$$

where

$\Psi(N)$ = probability of household having N members;
N = number of persons per household; and
$\bar{N}$ = average number of persons per household minus one.

Estimates of the average number of persons per household for any small geographical area allow the distribution of households by size to be calculated.

A binomial distribution is used to represent the number of employed persons. The probability that W members of an N person household are employed is:

$$\Xi(W\|N) = \frac{N!}{W!(N-W)!}\,(\bar{W})^{W}\,(1-\bar{W})^{N-W} \qquad (4.16)$$

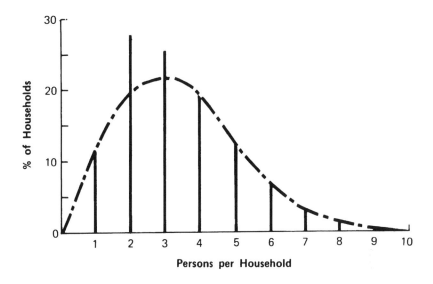

Figure 4.4: Category Analysis Family Size Distribution
Source: Based on Golding, 1972, Figure 4, p. 319.

where

$\Xi(W/N)$ = probability that W members of an N member household are employed;

$\bar{W}$ = probability that a member of an N-size household is employed, taken to be average work-force participation rate; and

W = number of employed persons per household.

Equations (4.15) and (4.16) are multiplied together:

$$\Pi(N,W) = \frac{N! \exp(-\bar{N})\,\bar{N}^{N-1}}{W!(N-W)!\,(N-1)!}\,(\bar{W})^{W}\,(1-\bar{W})^{N-W} \quad (4.17)$$

where

$\Pi(N,W)$ = probability that a household has N members of whom W are employed.

Equation (4.17) is expanded for appropriate values of family size and employed persons (Wilson, 1974, p. 139) to give the probability of a household being a specified category (see section 3.2.2).

4.3.4 Households Cross-Classified

The final synthesis is to construct the joint probability distribution of car ownership, income and household structure:

$$p(I,C,N,W) = \Pi(N,W) \int_{a_I}^{a_{I+1}} \Phi(C\|I)\,P(I)\mathrm{d}(I) \qquad (4.18)$$

where

$p(I,C,N,W)$ = joint probability of being in income group I, owning C cars and having a household structure N,W;

I = income group with upper and lower bounds given by a_{I+1} and a_I.

The income term uses the mean value of future incomes and the car ownership term uses the mean value of future incomes relative to changes in car prices. The total number of zonal households forecasted is multiplied by equation (4.18) to give an estimate of the number of households in each cross-classification.

4.4 Forecasts of Travel Demand

Future traffic is assumed to be a function of future land-use activity and future transport supply. Initially, traffic calculations for some future urban development pattern are made given that there is a very minimal change to transport supply because this will indicate the magnitude of future problems and reveal where new transport facilities are most urgently required. Transport planning aims to design a transport system to serve the future pattern of land uses. Because of the uncertainty surrounding land-use forecasts, alternative land-use plans and alternative transport plans should be examined for their traffic implications.

Figure 4.5 outlines the main steps in the traffic forecasting process. On the left-hand side of this diagram, *present* land-use and transport give rise to traffic demands which are quantified by systems models calibrated on base-year data. These models—primarily equations of traffic generation, traffic distribution, modal split and traffic assignment —represent the system interactions *for any given* inputs of land-use activity and transport supply. The right-hand side of Figure 4.5 shows that future land-use and transport are the input variables to the calibrated travel demand models, and the solutions to these equations are the traffic forecasts.

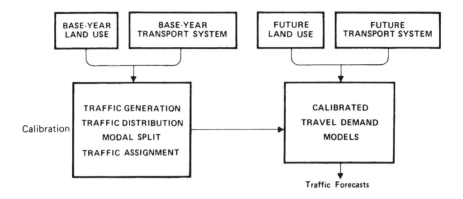

Figure 4.5: Traffic Forecasting Methodology

Transport planners are concerned most of all with the spatial and temporal dimension of traffic forecasts in the design of transport systems. The spatial detail is represented in the calibrated travel demand models by means of zonal labels. The main steps are listed below.

(a) Future zonal values of land-use activity are substituted into the traffic generation model to give forecasts of zonal traffic production and attraction.
(b) Forecasts of traffic production and attraction enter the traffic distribution model as variables, and the solution to these equations gives the estimated desire-line pattern of traffic.
(c) Modal-split and traffic assignment models convert this spatial pattern of traffic into mode- and route-specific flows based on the future operational characteristics of the transport system.
(d) The representation of future transport supply—network configuration and service levels—enters into the models as inter-zonal transport impedances.
(e) Because of the importance of transport supply on traffic there should be a feedback loop joining the models (Blunden, 1971, pp. 24–5) which ensures that the models are internally consistent and give an equilibrium solution.

Traffic forecasts are based on four important assumptions. First, the complex interactions between traffic, land-use and transport are correctly specified in the systems models. Second, the parameters of these models calibrated on cross-sectional data are stable through the forecasting period. Third, the future values of the land-use and transport variables which enter into the equations are forecasted accurately. Fourth, the effects of transport on land-use patterns are given little attention.

Traffic generation forecasts require either the regression coefficients or the category mean trip rates to remain constant over time. A comparison of forecasts from both regression and category analysis traffic production models calibrated on 1958 data for Pittsburgh with the traffic that eventuated found that total traffic was forecasted accurately, traffic stratified by trip purpose was less accurate, and, significantly, the zone forecasts were inaccurate: 'The degree of error found in these models would indicate that serious errors are possible over a longer and more usual planning period' (Ashford and Holloway, 1972, p. 806).

Different conclusions were reached by Downes and Gyenes (1976), based on a household survey of home-based travel by all transport modes (including walking and cycling) conducted in Reading in the autumn of 1962 and again in the autumn of 1971. Zonal forecasts for 1971 from the models calibrated on the 1962 data were just as reliable as the 1971 model estimates and therefore performed satisfactorily in forecasting traffic over a nine-year interval.

Structurally, the zonal regression equations for 1962 and 1971 differ in their explanatory variables and in their numerical values of the coefficients, and the household regression equations differ in their coefficients. Trip rates for certain identifiable groups within the population—especially retired persons, housewives and working women —altered significantly over time (Doubleday, 1977, Table 7, p. 260), thereby casting some doubt on the fundamental assumption underlying category analysis forecasts.

When forecasting with the gravity model the assumption is that either the calibration parameter in the transport impedance function or the 'friction factors' remain constant over time, although this has been challenged. Calibrated gravity models for different cities indicate that there are 'significant relations between curve parameters and such ephemeral variables as number of trips made, trip-making rates, car-ownership, and ratios of trips made for various purposes' (Ashford and Covault, 1969, p. 30) with the inference that the parameters are not constant, but are likely to change as the character of the urban area itself is modified over time.

Modal-split diversion curves for the journey to work in Toronto, calibrated for 1954 and 1964 data (Hill and Dodd, 1966), should theoretically produce an identical set of curves. Although the authors argue that the basic relationships developed from the 1954 survey data were 'still applicable in the planning process' (ibid., p. 19) there is little visual correspondence between the 1954 and 1964 curves.

Certain inconsistencies in modal-split modelling have serious implications for forecasting accuracy (Senior and Williams, 1977). Using data from the West Yorkshire conurbation, five different types of model were calibrated, but all gave different forecasts for identical transport system changes.

As suitable time-series data becomes available, further research may establish the degree of confidence that should be placed in traffic forecasts. This issue of forecasting inaccuracy is largely avoided if the model forecasts are interpreted as the *best estimates* available and are used to compare the traffic implications of alternative plans. In this way, alternative plans are evaluated on a consistent (although not necessarily on an entirely accurate) basis.

4.5 Transport Plans and Planning

Traffic forecasts are used by planners as a basis for justifying plans for urban transport facilities. In formulating transport plans, three types of plan may be usefully distinguished:

(a) long-term plans for the whole of urban areas that involve the construction of entirely *new* transport infrastructure;
(b) short-term action plans that involve improvements, alterations and modifications to parts of the *existing* transport system; and
(c) street layouts in new residential subdivisions.

Residential streets are considered in Chapter 9; this section explains how alternative transport plans are tested by systems modelling.

4.5.1 New Transport Facilities

Principles for the planning of major roads carrying heavy traffic volumes have emerged gradually over time (Tripp, 1942, p. 54; Gibberd, 1962, pp. 31–2; and Creighton, 1970, pp. 228–39):

(a) continuity–the major road network should be continuous

through an urban area, and not peter out or change direction too often;

(b) priority for moving vehicles—there should be no direct access from adjacent properties and buildings, no loading and unloading, no standing vehicles (all bus stops should be embayed) and complete segregation from pedestrians;

(c) equity—a fair geographical allocation of transport investment throughout the urban area; and

(d) high geometric design standards—intersections spaced ideally at a minimum of 1.5 km should be simple, and complex layouts with five (or more) approaches should be avoided.

There are no universal principles determining the ideal network configuration of main roads in each town. The traditional town planning approach is to consider only simple geometrical network configurations (Leibbrand, 1970, p. 91). The modern approach is to design a road system that conforms to the relative positions of the major traffic generators and the future desire-line pattern of traffic, and recognises the relevant topological constraints in each city (Witheford, 1976, p. 545).

In applying these principles planners must generate solutions tailored to the problems in their area; fortunately, the testing of alternatives is greatly aided by the computer. Additional links together with their capacities and expected transport impedances can be added to, or subtracted from, the base-year network representation of transport supply. Computer analysis is best illustrated with a simple example. Figure 4.6(a) shows a rectangular (two-way) network with labelled nodes and measured link travel times in minutes. As illustrated, the connectivity of this network and its operating performance is expressed entirely in numerical form, so it is suitable data for computer analysis. A proposed road plan involves linking nodes 2 and 3, and putting a new road on some parallel alignment to the existing road linking nodes 2 and 4. Figure 4.6(b) shows this network configuration, which together with its numerical description would replace the description of the old network in the computer analysis. The testing of any extensions to public transport are treated in a similar way.

4.5.2 Improving Existing Transport Facilities

An alternative approach to planning new facilities is to make better and more efficient use of the existing transport facilities. The road plans which are commonly proposed to improve transport are one-way

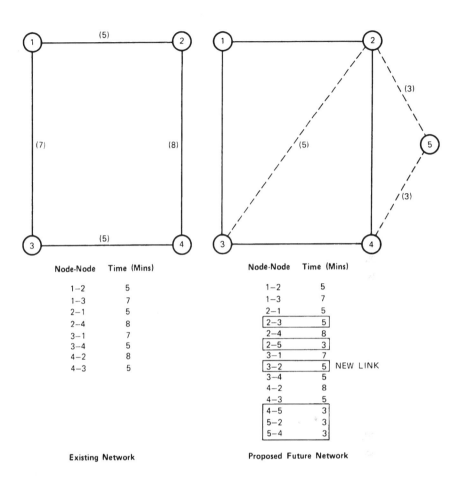

Node-Node	Time (Mins)
1–2	5
1–3	7
2–1	5
2–4	8
3–1	7
3–4	5
4–2	8
4–3	5

Existing Network

Node-Node	Time (Mins)	
1–2	5	
1–3	7	
2–1	5	
2–3	5	
2–4	8	
2–5	3	
3–1	7	
3–2	5	NEW LINK
3–4	5	
4–2	8	
4–3	5	
4–5	3	
5–2	3	
5–4	3	

Proposed Future Network

Figure 4.6: Transport Network Planning–New Roads

streets, traffic engineering measures such as co-ordinated signalised intersections, signs and markings, clearways and the closure of streets to through traffic; short-term public transport plans are more frequent services and revised fare structures.

The principles are explained with reference to the simple network presented earlier (Figure 4.6). The short-term plan to improve this network is to organise traffic flow in a clock-wise direction, thereby effecting a one-way street system and a revised network with half the number of links would be used in the computer analysis. Improvements to the operational performances of the road network with successful traffic engineering measures would be represented by lower link travel times. The street closure plan would involve removing a link from the network representation.

Better public transport operations are represented by reductions in link travel times, but revised fare structures can only be incorporated into the representation of future transport supply if generalised costs (see section 2.3.2) are specified as the link transport impedances. Figure 4.7(a) shows a residential zone centroid connected to a nearby railway station (node 2). The travel time from the centroid to the station is fifteen minutes as shown on the link connecting node 200 with node 1. The average waiting time for the train is represented by the 'dummy link' (node 1 to node 2). The plan is to introduce feeder bus services to and from the railway station and to increase the frequency of trains, and this is represented in Figure 4.7(b) with reduced access times and waiting times.

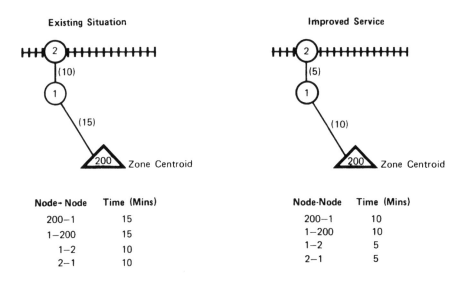

Node-Node	Time (Mins)
200−1	15
1−200	15
1−2	10
2−1	10

Node-Node	Time (Mins)
200−1	10
1−200	10
1−2	5
2−1	5

Figure 4.7: Transport Network Planning–Improved Public Transport Services

The purpose of representing alternative plans for future transport supply as networks of nodes, links and link impedances is to compare and contrast the assignment of forecast traffic to these networks. With the above simple examples, the traffic assignments can be done manually, but with realistic networks the computer is an invaluable aid to the planner. The criteria for choosing the best plan from a set of alternatives involves evaluation, which is considered in the next section.

4.6 Evaluation of Land-use–Transport Plans

When alternative land-use–transport plans are proposed a procedure is required to help judge which one is the preferable course of action. Basically, the evaluation method with urban transport investment is to calculate the additional benefits less implementation costs from each plan over and above the situation if no plan were implemented, and to recommend that plan if it has greater benefits than the 'do-nothing' situation. The comparative appraisal is carried out in an objective way, normally using the 'measuring rod of money' but, in more recent years, including factors which do not have a market value, such as pollution.

This cost-benefit analysis is detailed in some economics and public finance text books (for example, Mishan, 1971; Musgrave and Musgrave, 1973, pp. 109-66) and readers are recommended to consult them for the basic economic principles and assumptions. There are also books on the specific topic of transport project appraisal (for example, Harrison, 1974; Jones, 1977), which show the practical applications of cost-benefit analysis. Because the following explanation is necessarily brief, it should be supplemented with additional reading.

4.6.1 Costs and Benefits

Whilst the costs of implementing a transport project are straightforward conceptually, benefits to travellers need further explanation within a simple economic framework. Technically, user benefits are expressed as the 'consumer surplus', which is defined as 'the maximum a consumer will pay for a given amount of a good, less the amount he actually pays' (Mishan, 1971, p. 31). In the transport context, this is the difference between the cost of travel and the maximum amount that a traveller would be prepared to pay in completing a journey.

Figure 4.8 shows the demand curve for travel between two places, drawn here for simplicity as a straight line labelled AA', plotted against the transport cost of making the journey. (This diagram should be compared with Figures 1.5 and 1.6.) For clarity, the transport supply function has been omitted but the equilibrium point between transport supply and transport demand is indicated by point B. Hence, the equilibrium traffic demand is Q_1 and the transport cost at this traffic flow is T_1. According to the complete demand curve, some travellers enjoy a consumer surplus–the area of the triangle ABT_1.

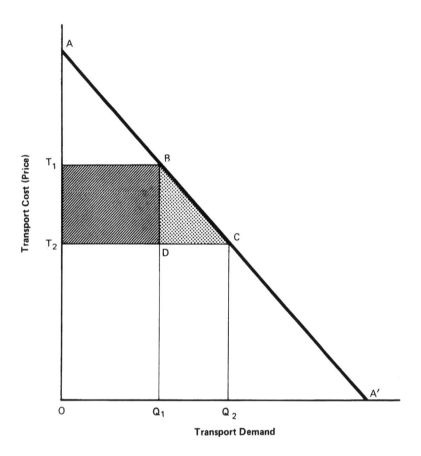

Figure 4.8: Measure of Consumer Surplus

Consider a transport plan which is expected to alter the shape of the transport supply function, thereby producing a new equilibrium point C in Figure 4.8. Transport costs are lowered to T_2 which induces some more trips to be made, thereby increasing the demand for transport to traffic flow Q_2. As far as the original users are concerned, they benefit from an increase in consumer surplus which corresponds to the darker shaded rectangle BDT_2T_1. Benefits to the new users are represented by the stippled triangle *BCD*. Therefore, the total benefits of the transport plan are represented by the shaded trapezium, which can be calculated algebraically as the *change* in consumer surplus:

$$\Delta CS = \tfrac{1}{2}(Q_1 + Q_2)(T_1 - T_2) \qquad (4.19)$$

where

ΔCS = change in consumer surplus;
Q_1 = transport demand before the plan;
Q_2 = transport demand after the plan;
T_1 = transport cost before the plan; and
T_2 = transport cost after the plan.

This explanation of the user benefit to people travelling between two places is applicable also for system-wide benefits, and equation (4.19) may be extended for all zone pairs and for all transport modes:

$$\Delta CS = \sum_i \sum_j \sum_m \tfrac{1}{2}(Q_{ij(m)} + \hat{Q}_{ij(m)})(T_{ij(m)} - \hat{T}_{ij(m)}) \qquad (4.20)$$

where

ΔCS = change in consumer surplus;
$Q_{ij(m)}$ = base-year traffic by transport mode m from zone i to zone j;
$\hat{Q}_{ij(m)}$ = forecast traffic by transport mode m from zone i to zone j, after the plan;
$T_{ij(m)}$ = base-year transport cost by transport mode m from zone i to zone j; and
$\hat{T}_{ij(m)}$ = estimated transport cost by transport mode m from zone i to zone j, after the plan.

Whereas the major benefit of a proposed urban transport plan would be the reduction in transport impedance with its associated imputed monetary values, the costs of the plan are the market prices of acquiring the necessary land, or property, and the cost of the labour and material required to construct and operate the new transport system. In principle, it is simpler to disregard the distinction between capital costs and operating costs and enter all payments as costs. The main difficulty in measuring costs correctly is that for some items market prices may be inappropriate. For instance, in times of high unemployment in the private construction industry governments may tap this source of labour for public works projects, and the prevailing market wage should be modified to take into account the social security payments that the unemployed would otherwise receive.

4.6.2 Evaluation Criteria

Practical difficulties arise in the identification, specification and accurate measurement of *all* relevant costs and benefits, not least because of the enormous and wide-ranging impact of transport investment on society. Major projects have a lengthy gestation period and usually continue to produce benefits far into the future, so there is the additional complication of the temporal profile of costs and benefits. Assuming that the costs and benefits can be measured properly for each alternative plan, and their expected time profile established, the question is which plan should be recommended.

Meaningful answers are obtained only if the net benefits of each alternative are comparable. Differences in the time stream of net benefits are overcome by discounting them with the appropriate interest rate, which reflects society's valuation of future benefits as compared with benefits available now (Hotchkiss, 1977, p. 45). In many countries, governments specify appropriate discount rates for public sector investment. The yield on government bonds—the market rate of interest—is generally regarded as a sufficiently good indication of the 'social rate of time preference', but it does have its critics (Mishan, 1971, pp. 202–14).

Once an appropriate discount rate has been chosen, the problem of comparing annual costs and benefits over time largely reduces to a matter of arithmetic. The three investment criteria are (a) the net present value, (b) the internal rate of return, and (c) the benefit-cost ratio. The formula to calculate the net present discounted value of an investment proposal is:

$$NPV = \sum_{t=0}^{t=n} (B_t - C_t)/(1 + \rho)^t \qquad (4.21)$$

where
NPV = net present value;
B_t = benefits in year t;
C_t = costs in year t;
ρ = discount rate, and
t = time in years, starting with 0 as the initial year.

Any plan is worthwhile if the net present value is greater than zero. When evaluating alternative plans the criterion is to choose the plan that gives the highest positive net present value.

The calculation of the internal rate of return involves the above formula: net present value is set to zero and the value of the 'unknown' discount rate ρ, which satisfies the equality, is the internal rate of return. In comparing alternatives, the criterion is to choose the plan with the highest internal rate of return. Provided that the internal rate of return exceeds the adopted discount rate, any public investment programme is justified economically.

As implied by its name, the benefit-cost ratio is a straightforward calculation involving the discounted stream of benefits and the discounted stream of costs. Despite theoretical objections to its use in public sector investment (Harrison, 1974, pp. 200-2), benefit-cost ratios indicate clearly when a plan is near the break-even point of unity. In such circumstances, any mis-specifications in costs and benefits may switch the plan to the other side of the break-even point.

Because accurate measurements of all relevant costs and benefits are difficult to obtain sensitivity analyses are an essential part of cost-benefit analysis. Measurements or assumptions about the costs and benefits are varied in a controlled way to assess the sensitivity of the final result to these systematic changes. The aim is to investigate whether the ranking of alternative plans remains the same or not.

A simple example demonstrates how straightforward the computations are to obtain the net present value, the internal rate of return, and the benefit-cost ratio. The discount rate is 10 per cent per annum and the time stream of benefits and costs are:

Year	t	0	1	2	3
Annual benefits (\$)	B_t	0	500	3,000	3,000
Annual costs (\$)	C_t	3,000	1,000	1,000	0

From equation (4.21):

$$NPV = \$452.3 \text{ (one decimal place).}$$

The internal rate of return is obtained from an iterative procedure by assuming an initial value for the discount rate in the above calculation and revising this until the net present value equals zero. The internal rate of return is 15.7 per cent per annum (correct to one decimal place). Finally, the discounted stream of benefits (B) is:

$$B = \$5,187.8 \text{ (one decimal place).}$$

The discounted stream of costs (C) is:

$$C = \$4,735.5 \text{ (one decimal place).}$$

Therefore the benefit-cost ratio (B/C) is 1.1, correct to one decimal place. The sensitivity of the three answers to changes in the benefits, costs or discount rates will become apparent if the reader re-works the calculation with a 15 per cent discount rate *and* an increase of 10 per cent in costs $(NPV = -\$412.5; \rho = 9.8\%$ per annum; $B/C = 0.9)$.

4.6.3 Environmental and Social Factors

Ideally, the scope of evaluation should broaden to include factors which may have no obvious monetary dimension but which nevertheless are relevant in the appraisal of alternative plans. Large civil engineering works in urban areas have environmental repercussions which affect the welfare of residents and should be included in any comprehensive evaluation. During the construction of new transport facilities, noise, dust and fumes inconvenience pedestrians, and nearby residents and workers alike; once complete and operational, there are problems of vehicle noise, vibrations and emissions. A good general introduction to this complex area is *Transport and the Environment* (Sharp and Jennings, 1976). Whilst it is relatively easy to make a list of the environmental consequences of transport investment, it is very difficult to predict their magnitude, especially since the impacts are so location-specific. Noise prediction models are becoming part of standard practice (DOE, 1975) but as yet there are no equivalent models to predict reliably atmospheric pollution caused by vehicles.

A framework for setting out economic, environmental and social benefits and costs of any plan is provided by the *planning balance sheet* (Lichfield, 1966). Individuals, groups or government agencies who play a part in creating and running transport facilities are listed at the left-hand side of the balance sheet, and each 'producer or operator' is paired with an appropriate 'consumer' listed on the right-hand side of the balance sheet. Each linked pair are considered engaged in a transaction–a 'flow' of costs or benefits, which may include non-quantifiable social or environmental transactions. When complete, the table is a set of 'social accounts' summarising all transactions and the number of people affected where known. The advantage of this format is that double counting is avoided and a comprehensive picture of the incidence of transactions on all relevant parties is obtained. Unfortunately, such a presentation does not lend itself to clear-cut statements of the desirability or otherwise of the project and value judgements 'involving all the classical problems of social choice' are present (ibid., p. 241).

The distributional implications of plans become increasingly relevant when a broader social evaluation is attempted. One way of incorporating the spatial distribution of benefits from alternative land-use and transport arrangements is to measure the zonal accessibility properties of each plan (Neuburger, 1971; Black and Conroy, 1977; Conroy, 1978).

4.7 Summary

The systems approach suggests that once the fundamental interactions in the system of interest have been modelled then the future state of the system can be predicted. Traffic forecasts are calculated from assumptions about future land-use and this chapter has described methods of projecting population, employment, income and car ownership at the aggregate and disaggregate scale. Planning involves the consideration of alternative land-use and transport arrangements. The testing of alternative transport plans, both for modifying the existing roads and public transport services or for extending the system, can be carried out by altering transport supply in the systems model in an appropriate way.

Forecasts of the traffic implications of different land-use—transport plans lead to the next step in the systems approach: the evaluation of alternatives. The economic approach to transport appraisal was discussed, and three investment criteria were presented: the net present value, the internal rate of return and the benefit-cost ratio. Social and environmental factors are also important but are more difficult to include into the appraisal.

Examples of urban land-use—transport planning are given in the next two chapters and the forecasting methodology employed should be compared and contrasted with the theory presented in this chapter. Reference is made to evaluation methods in practice in the next four chapters, but environmental considerations appear only in the appraisal of public transport technologies (see section 7.3).

PART TWO: PRACTICE

What was done by heart and will
is now being done by the head.
Political faction and fantasies of design
have been succeeded by social and economic analysis
and the aid of computers;
so that the planners understand
a great deal more than they did
of the choices and possibilities before them,
and the public are made aware of the options
and can help in making the decisions
that affect the future of the National Capital

Sir William Holford (1972, p. 27).

5   CONVENTIONAL LAND-USE–TRANSPORT PLANNING STUDIES

Land-use–transport planning studies are designed to help governments to plan transport investment in urban areas more effectively. Such studies, conducted according to more or less standard procedures and often taking three years to complete (Sweet, 1969, p. 26), have been carried out in most major cities throughout the world (Witheford, 1976; Thomson, 1977), and in smaller ones–hence the label 'conventional' in this chapter heading.

Figure 5.1 shows the major steps in the urban transport planning process, which forms the backbone to all major transport studies conducted throughout the world. This remarkably uniform methodology has prompted Ben Bouanah and Stein (1978) to suggest there is 'a generalised international urban transportation planning process'. The technical procedures, first pioneered in the Detroit and Chicago transportation studies of the mid-1950s, have been vigorously promoted by government agencies (BPR, 1964, 1965 and 1967), and by a few major engineering and planning consultant organisations.

Because of the similarity in the methodology of transport studies, it is instructive to examine a case study. The goals and objectives, data collection procedures, travel demand modelling, land-use and traffic forecasting, transport network planning and evaluation procedures adopted in planning for land-use and transport in Canberra are described. This chapter illustrates aspects of the theory explained in Part One.

5.1 Canberra Area Transportation Study Objectives

Although the Commonwealth government had launched a competition in 1911 for laying out the new federal capital and had adopted Walter Burley Griffin's design which became the Gazetted Plan of 1925 (Linge, 1975), Canberra failed to develop satisfactorily as a national capital. The government recognised that the administrative arrangements for urban development, with responsibilities divided amongst departments, were deficient, and established, in 1957, the National Capital Development ment Commission (NCDC) as a statutory body 'to undertake and carry

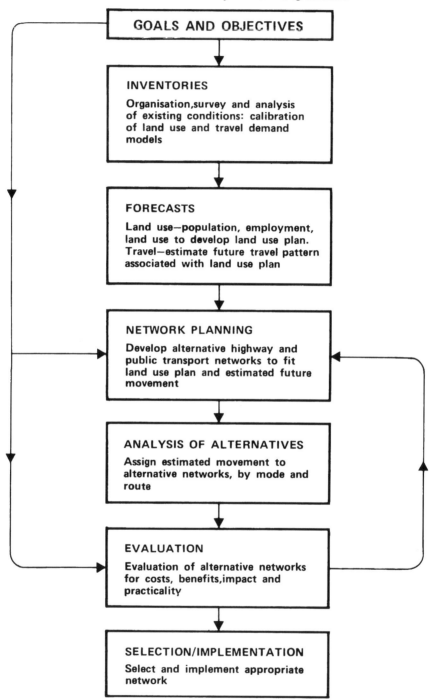

Figure 5.1: Land-use–Transport Planning Process
Source: Based on BPR, 1970.

out the planning, development and construction of the City of Canberra as the National Capital'.

Transport planning was one of the important components in the preparation of a Development Plan. In 1958, Canberra's population was 39,000 and had doubled during the previous seven years. The number of vehicles had trebled and the number of traffic accidents quadrupled in the same period. The primary objective of the *Canberra Area Transportation Study Engineering Report* (CATS, 1963, p. 6) was 'to develop a transportation system which would efficiently and economically accommodate the transportation needs for Canberra up to a population of 250,000'. On the then current growth forecasts the population was expected to reach a quarter of a million by the end of the century.

A continuing dependency on the motor car was assumed, although public transport was examined also to provide a 'balanced' transport system. An important feature of the Griffin plan is the central area composition, covering about 30 square kilometres, and suitable traffic arrangements were required.

5.2 Surveys and Analysis of Travel Demand

The study area was defined as the built-up parts of Canberra plus the surrounding countryside, designated as being environmentally suitable to accommodate future urban development. Figure 5.2 shows the delineation of 107 zones in the study area. The only suitable land-use data available were population, employment and vehicle ownership. The operating characteristics of the transport system were established from travel time surveys of roads and public transport. Information on the travel patterns in Canberra was established from postcard surveys of owners of registered vehicles and of passengers on the buses. Roadside surveys provided information about traffic with origins external to the study area.

Person travel in Canberra was represented by models of traffic production, traffic distribution, and traffic assignment. There was no modelling of traffic attraction and assumptions were made about the modal split between private and public transport.

The statistical technique used to model zonal traffic production was simple linear regression (section 3.2.1). The dependent variable was defined as the zonal average number of daily person-trips per resident leaving the zone (i.e. one-way travel) and was stratified into

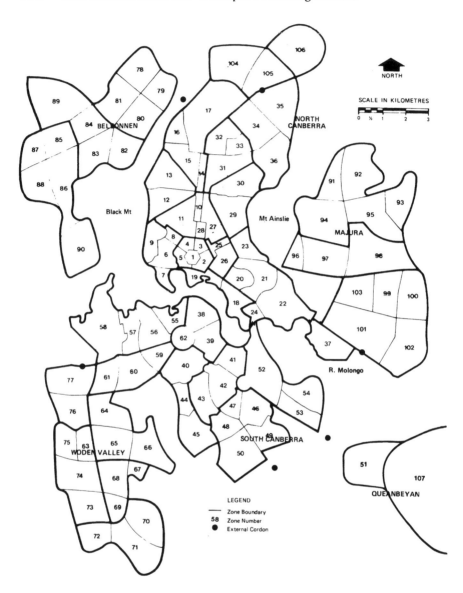

Figure 5.2: Canberra Area Transportation Study Zone System
Source: Based on CATS, 1963, Exhibit 2.

four journey purposes. The parameters are given in Table 5.1. The general form of the traffic production model (cf. equation (3.1.)) is:

$$\hat{Q}^r_{pi} = \alpha + \beta(P_i/V_i)$$

where

$\hat{Q}^r_{pi}$ = estimate of the daily number of trips by purpose r leaving zone i;

P_i = population of zone i; and

V_i = vehicle registrations in zone i.

Zonal totals for traffic production were calculated by multiplying the trip estimates by zonal population. Non-home-based traffic production was obtained by multiplying trip estimates by a weighted combination zonal population (10 per cent) and zonal employment (90 per cent).

Table 5.1: Parameters of the Zonal Traffic Production Model, Canberra, 1961

Trip Purpose (r)	Constant α	Regression Coefficient β
1. Home-based work	+0.67	−0.08
2. Home-based business/shopping	+0.43	−0.05
3. Home-based social/recreation	+0.42	−0.07
4. Non-home-based	+1.84	−0.19

Source: CATS, 1963, pp. 68–9.

A production-constrained gravity model with a power function of travel time (equation (3.9)) was used to estimate the spatial pattern of traffic. Table 5.2 gives the calibration parameters for each journey purpose (see section 3.3.2). Because there was no modelling of traffic attraction equation (3.9) contains a measure of land-use intensity for the jth zone, as indicated in Table 5.2.

Table 5.2: Parameters of the Traffic Distribution Model, Canberra, 1961

Trip Purpose (r)	Attraction Measure (L^r_{dj})	Parameter (α)
1. Home-based work	Total employment	0.525
2. Home-based business/shopping	Commercial employment	2.206
3. Home-based social/recreation	Population	0.528
4. Non-home-based	Total employment	0.696

Source: CATS, 1963, pp. 69–71.

There was no modelling of mode choice and the survey modal split of 7 per cent of daily person travel by bus and 13 per cent of peak period travel by bus was assumed to remain the same. The number of person trips by private transport was converted by applying average vehicle occupancy rates by journey purpose.

Passengers were allocated to the public transport network by an *all-or-nothing* assignment (section 3.5.1). Vehicular traffic was allocated to alternative routes by an empirically drawn diversion curve that shows the proportion of traffic using the best route against the ratio of the travel times on that route compared with the travel times along the next best route (BPR, 1960).

5.3 Land-use and Traffic Forecasts

The future use of land in the Canberra region was specified by the NCDC planners. Projections of population, employment and vehicle registrations were made for each of the 107 zones. Because the NCDC can determine the location of land uses and can control residential densities (section 4.2), forecasts of population and employment for small areal units are likely to be reasonably reliable. Projections of vehicle ownership were an extrapolation of national trends (section 4.1.3), and are more uncertain in the long term.

Traffic forecasts were made by substituting these zonal 'land-use' projections into the calibrated models of traffic generation, traffic distribution and traffic assignment, given assumptions about modal split. Figure 5.3 illustrates the results of the first three stages of traffic forecasting by showing the origin-destination pattern of daily vehicular traffic aggregated for all trip purposes. Figure 5.4 gives an example of one of the traffic assignments: the volumes shown represent daily two-way traffic on the recommended highway network.

5.4 Network Planning, Evaluation and Implementation

Network planning involved testing different configurations for highways and public transport, and specifying highway design standards. The recommended highway network for Canberra at the 250,000 population level contains 660 links. These links were then dimensioned by the number of carriageway lanes required to accommodate the traffic. Roads were designed to handle peak-hour traffic which was obtained from factoring daily traffic.

The saturation flow of expressways and arterial roads were: 1,500 vehicles per hour per lane of expressway; and 450 vehicles per hour per lane of arterial, assuming 50 per cent green-time at signalised intersections (CATS, 1963, Table 11).

*Figure 5.3: Daily Vehicle Traffic Desire Lines, Canberra, at the 250,000
Population Level*
Source: Based on CATS, 1963, Exhibit 5.

Two alternative public transport systems were tested: a railway with
feeder-bus services, and a modification to existing bus services. The
proposed railway is a 10 km route from Crace in the north to Fyshwick
in the south-east. Various options within the broad framework of
altering existing bus services were examined, including express buses

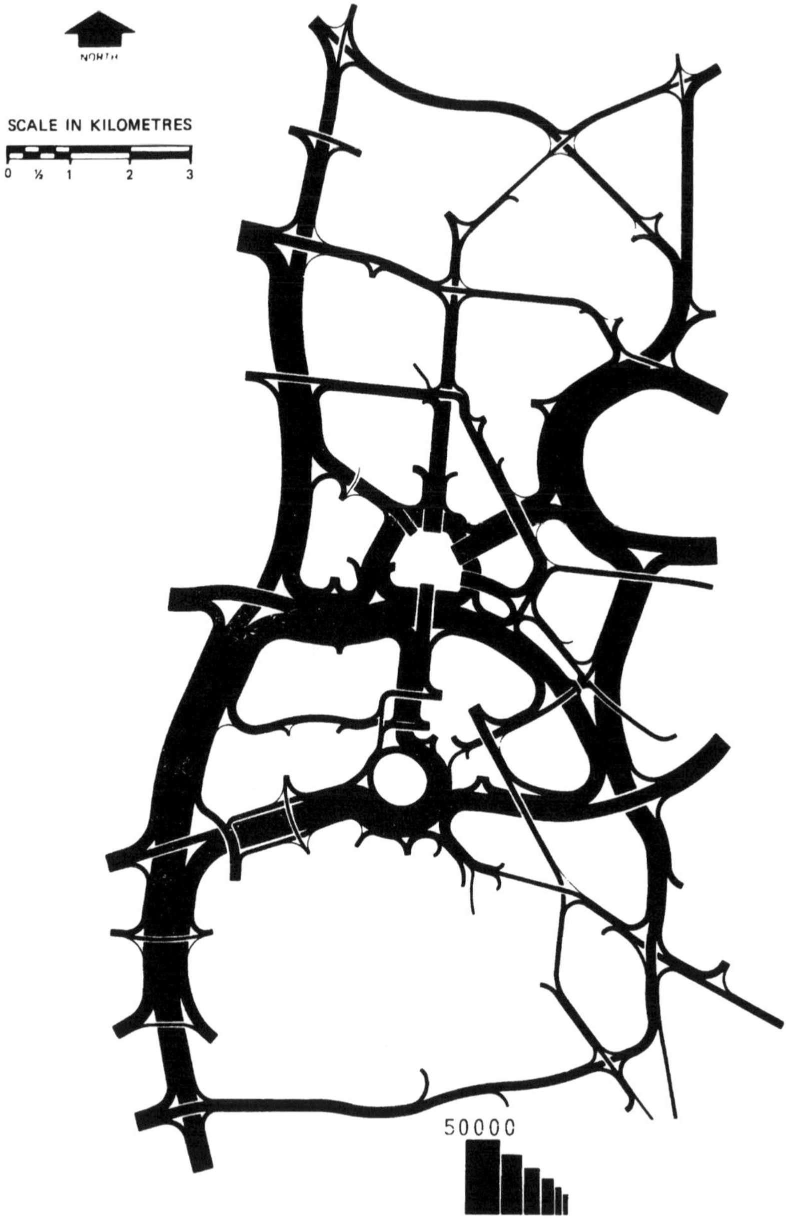

Figure 5.4: Daily Traffic Assignment, Canberra, at the 250,000 Population Level
Source: Based on CATS, 1963, Exhibit 7.

supported by feeder-buses radiating from each town centre, existing services with peak hour express buses to and from major employment centres, and the existing system but with restructured and more direct routes.

The recommended transport plan was based on the suitability of the facilities to accommodate future traffic and the costs of providing these facilities. No economic nor environmental evaluations were made of the alternative plans. Cost estimates for the road plan were based on the unit rates in contracts tendered by the NCDC. The total cost of the plan was £A43.6 million (1963 prices), excluding the extra cost of property resumption, compensation and special landscaping.

The capital cost of the railway, including rolling stock, was estimated at £A22 million (1963 prices). Operating, maintenance and equipment replacement would cost an extra £A0.6 million each year. The consultants did not recommend the railway and 'considered that the present framework of bus services, suitably modified from time to time to meet the emerging needs of the City, affords the best basis for the development of future public transport' (CATS, 1963, p. 38).

The NCDC did not adopt the recommendations. The centralised location of employment and dispersed pattern of residences would generate a complex set of traffic desire lines (Figure 5.3) and the highways needed to cope with the traffic forecasted would lower residential amenity and destroy the architectural composition of the central area, which is the heart of Griffin's plan. The hexagonal express-way system surrounding Civic Centre together with its east-west distributors would be highly intrusive.

The methodological weakness of the *Canberra Area Transportation Study* was its failure to examine the traffic implications of other urban development strategies. It was only when the final traffic assignments were available that the NCDC planners fully appreciated the traffic implications of their original land-use plan. The planners looked for a better plan to guide long-term urban development and this story is continued in Chapter 6.

5.5 Continuing Transport Planning

Urban transport planning is not a 'one-off' exercise leading to an inflexible master plan, but is a continuing process (Witheford, 1976, p. 516). In Canberra, the ability to do this was enhanced recently with the *Canberra Short Term Transport Planning Study*, an objective of which was to establish an in-house transport modelling 'capability to operate and interpret the models and methods developed in order to continue the study's analysis in the future' (CSTTPS, 1977a, p. 1).

The travel demand model comprised of category analysis for traffic production, trip rates by land-use classification for traffic attraction, a combined traffic distribution–modal-split model, and a capacity restraint traffic assignment model. The model was calibrated on data collected from a home interview survey at 2,253 dwellings. The study area was partitioned into 165 zones, and person travel was stratified into six journey purpose categories: home-based work, shopping, education and other; non-home-based shopping; and non-home-based other (CSTTPS, 1976).

Households were cross-classified by six categories according to the number of workers and non-workers and by three car ownership categories (cf. section 3.2.2), but the very small sample meant that some categories were combined, as shown in Table 5.3. The category mean trip rate (including walking and cycling) for daily, two-way person travel, are also given although the transport study gives more detailed tabulations.

Table 5.3: Daily Mean Trip Rates for Households, Canberra, 1976

Number of Workers		0	.	1	.	≥2	.
Number of Non-Workers		≤2	≥3	≤2	≥3	≤2	≥3
Number of Motor cars — 0		3.3		4.0		7.9	
Number of Motor cars — 1				6.2	10.1	9.3	16.1
Number of Motor cars		5.0	10.7				
Number of Motor cars — ≥2				8.1	12.0	11.7	16.6

Source: Based on CSTTPS, 1977b, Table 2, p. 3.

The calibration parameters of the combined traffic distribution–modal-split model (section 3.4.1) are given in Table 5.4. Separate models were specified for home-based work, home-based education, home-based other and non-home-based travel. The general form of the model is:

$$\hat{Q}^{rs}_{ij(m)} = \frac{Q^{rs}_{pi}Q^{rs}_{aj}/\exp(\beta^{rs}_m T_{ij(m)} + \delta_m)}{\displaystyle\sum_{j=1}^{165}\sum_{m=1}^{3} Q^{rs}_{aj}/\exp(\beta^{rs}_m T_{ij(m)} + \delta_m)}$$

where

$\hat{Q}^{rs}_{ij(m)}$ = estimate of the mode-specific m daily traffic from zone i to zone j for journey purpose r by person-type s;

Q^{rs}_{pi} = traffic production of zone i for journey purpose r and person-type s;

Q^{rs}_{aj} = traffic attraction to zone j for journey purpose r and person-type s;

$T_{ij(m)}$ = generalised cost of travel from zone i to zone j by transport mode m; and

δ_m = mode-specific handicap penalty.

The three transport modes are car driver ($m=1$), car passenger ($m=2$) and bus passenger ($m=3$), and the two person-types are people from households with no motor car ($s=0$) and people from households with one (or more) motor cars ($s=1$).

Table 5.4: Parameters of the Traffic Distribution–Modal-split Model, Canberra, 1976

Journey Purpose	Parameter	Transport Mode; Person-type Code				
		2;0	3;0	1;1	2;1	3;1
Work	β	0.115	0.028	0.051	0.115	0.028
	δ	0.769	0	0	1.467	0.688
Education	β	0.413	0.031	0.057	0.413	0.031
	δ	0	1.503	0	-2.574	-1.512
Other	β	0.191	0.047	0.108	0.191	0.047
	δ	0	0.120	0	1.249	1.184
NHB	β	–	–	0.070	0.169	0.045
	δ	–	–	0	0.958	0.518

Transport Mode Code	Person-type Code
1 – car driver	0 – person from zero car household
2 – car passenger	1 – person from car owning household
3 – bus passenger	

Source: Based on CSTTPS, 1977b, Table 6, p. 22.

Finally, the traffic assignment model is the Urban Transport Planning System (UTPS) computer software package. An *all-or-nothing* assignment is used to allocate passengers to the public transport network. A capacity restraint assignment (section 3.5.1) allocates vehicular trips to the road network. The Bureau of Public Roads formula (equation (2.17)) is used for the traffic flow-dependent travel times.

5.6 Summary

The systems approach has been used as a framework for organising the description of the steps of the *Canberra Area Transportation Study*. Although there are aspects of this study which are peculiar to Canberra, there are many features in common with land-use–transport studies conducted during the 1960s in cities throughout the world, not least the failure to examine alternative land-use plans and to give proper consideration to public transport. The planning studies described in the next two chapters largely overcome such criticisms.

6 LONG-TERM, STRATEGIC TRANSPORT PLANNING

When cities are expanding rapidly, or when decisions are taken to build new towns, planners are confronted with numerous possibilities for accommodating long-term growth. Strategic transport planning examines the traffic implications of alternative land-use options and recommends the best pattern and staging of development. Because of the necessity of considering many land-use—transport concepts the strategic planning process differs from the conventional transport planning process described in the previous chapter. An important distinction is the use of simplified urban travel demand models that reduce the cost and complexity of analysing options.

To begin with, the strategic transport planning process is defined and its methodology is outlined; this should clarify the difference between the conventional approach described in the previous chapter and the approach taken in this chapter. An important part of strategic planning is the more simplified travel demand analysis, which is explained in section 6.2 with examples. Later sections describe the *Canberra Land Use Transportation Study* (1967), which provides an instructive case study of this type of planning.

6.1 Strategic Land-use—Transport Planning

The principal purpose of long-term urban strategy planning, or 'sketch planning' as it is sometimes called in North America, is to examine different urban development concepts in sufficient detail to allow the traffic and transport implications of each alternative to be analysed. Strategic urban transport planning is the process of determining a recommended long-range level of investment in transport, dividing this investment amongst the main transport modes, identifying future corridors for road and public transport facilities, and suggesting the best timing for investment. When examining radically different future patterns of urbanisation, or different policies, it is necessary to look some twenty or thirty years ahead and to consider the broad outcomes rather than the details. A more thorough transport systems planning study may follow on from the most promising or feasible alternatives (Hutchinson, 1974, p. 333).

147

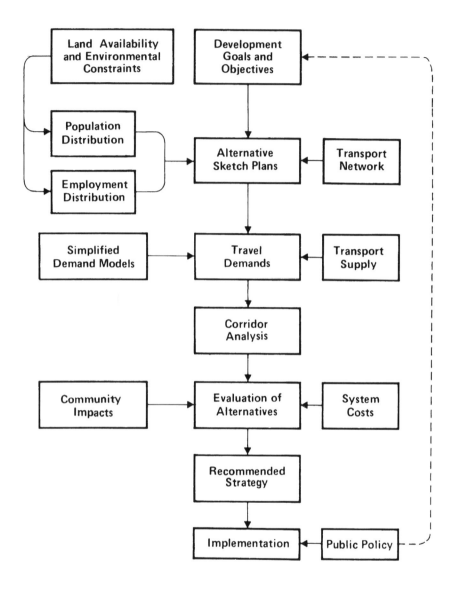

Figure 6.1: Strategic Land-use–Transport Planning Process

Figure 6.1 outlines the main components of a long-term, strategic
land-use–transport planning process. Urban development objectives
and policies, together with forecasts of population and employment,
and constraints imposed by land availability and other environmental
factors, lead to the formulation of alternative sketch plans. A transport
network is specified and future travel demands are calculated from the

interaction of land-use activity and transport supply. The traffic implications are assessed on a corridor-by-corridor basis—either a major route or a series of parallel routes in a transport corridor. Alternative land-use—transport alternatives are evaluated by their impact on the community and on the environment and by their implementation costs.

The general aims of strategic planning are: (a) to minimise the adverse effects of long-term growth on the existing urban fabric; (b) to minimise the effects of future urban development on the natural environment; (c) to contain journey-to-work trip lengths by a sensible balance of residential and employment opportunities; and (d) to develop a financially feasible transport system that is compatible with the environment, and with the preferences of the community. Although these aims may be shared by conventional transport studies (see Chapter 5), strategic planning is distinguished by an appropriate methodology that is simple, fast and inexpensive for examining the traffic implications of scores of land-use—transport alternatives and their phasings. As noted in Figure 6.1, simplified analytical techniques are one way of obtaining the equilibrium between transport supply and travel demand.

6.2 Simplified Urban Travel Demand Analysis

There are two practical reasons for developing special analytical approaches for strategic planning purposes. First, the cost of examining a large number of alternatives may be prohibitive when using conventional models. For example, the cost at the more detailed systems planning level can be from ten to twenty times that at the strategic level (Dial, 1976, p. 45). Second, many transport planning authorities need a method of appraising strategic policies without detailed studies involving data collection and the running of complex and costly mathematical models (OECD, 1974, p. 7).

Some suitable approaches towards analytical simplification include:

(a) large zones, and/or sparse transport networks;
(b) special-purpose computer programmes to reduce computational times and costs;
(c) restricted number of trip-purpose categories, with emphasis on the journey to work;
(d) broad trends and average values based on comparative studies of urban areas; and

(e) models developed and calibrated in one city which are transferable to another city.

Large zones and sparse networks reduce costs because the analysis
is conducted for a small number of macro-zones, and transport supply
is confined to major roads—such as motorways, expressways, arterial
roads and the line-haul parts of the public transport system. Conventional
travel demand models are still used. Examples of this approach are the
sketch plan option in the methodology developed by Cambridge
Systematics for the US Department of Transportation (DOT, 1974),
and the circular, symmetrical network of the CRISTAL model developed
by the Transport and Road Research Laboratory (Tanner *et al.,* 1973).

An example of a special computer programme is COMPACT, recommended for the initial sifting of broad strategies and for structure plan
analysis (Mackinder, 1972). The mathematical modelling techniques
are the same as those used in conventional transport studies, but the
number of zones are restricted to 98 or fewer and the transport
network description is confined to 540 links or fewer, and 135 nodes
(including centroids). Consequently, all matrices required in the
computations may be stored and operated on within the central
memory of the computer and costs are cut because the expensive
transfer of data to and from 'backing store' is minimised. (A
commercially available version called ENCORE allows the analysis of
up to 100 zones, 400 nodes and 1,000 links.)

Limiting the analysis to the journey to work or, at the most, to a
trip stratification of work and 'other' purposes, recognises that, for
strategic planning purposes, travel between home and work is the most
important. Investigations of the rush-hour traffic are usually sufficient
because a transport system that caters for the peak invariably caters
for off-peak travel too. A good example of this approach would be the
Lowry-type land-use—transport model (Hutchinson, 1974, Chapter 6).

The wealth of data collected during the course of conducting
conventional transport studies allows broad trends in urban travel
demand over time, and average values of key parameters or variables
to be computed for cities of different size. This is one approach towards
eliminating costly base-year surveys and avoiding the lengthy process
of model calibration. *Characteristics of Urban Transportation Demand*
(Wilbur Smith *et al.,* 1977) is a compendium of information on urban
travel behaviour and transport system usage in US cities. The manual
was designed to indicate key parameters for travel demand analyses,
and to simplify procedures for estimating or verifying urban demands
(Levinson, 1978, p. 2).

These guidelines should be read in conjunction with *Transportation and Parking for Tomorrow's Cities* (Wilbur Smith *et al.*, 1966, Appendix A), which suggests how to identify the required amount of CBD parking, and the required length of freeways in cities of differing sizes and characteristics. The ranges given are 'a point of departure—a *first* attempt at scaling the basic elements of the urban transport plan' (ibid., p. 295). Parking in the CBD is a function of city size and the expected number of vehicle journeys each day. The length of urban freeways is a function of city size and population density.

Trends in trip lengths are helpful in calculating the total amount of travel. Figure 6.2 shows the cross-sectional relationship between the mean trip length for the journey to work and population size for selected North American cities. Mean trip lengths in miles (Figure 6.2(a)) and the mean duration of travel by car (Figure 6.2(b)) generally increase by city size, although there is considerable variation for cities of similar size. Regression analysis gives the following general relationship between mean trip length in kilometres and urban population (Voorhees *et al.*, 1968, p. 9):

$$\bar{D} = 0.74 \, P^{0.19} \qquad (r^2 = 0.75)$$

An analysis of journey-to-work census data for Canadian cities Hutchinson and Smith, 1978) gave the following regression equation:

$$\bar{D} = 0.29 \, P^{0.25} \qquad (r^2 = 0.60)$$

The relationship between the mean travel time in minutes for the journey to work by car in the selected North American cities (Voorhees *et al.*, 1968, p. 9) is:

$$\bar{T} = 0.98 \, P^{0.19} \qquad (r^2 = 0.71)$$

where
$\bar{D}$ = average distance travelled to work in kilometres;
$\bar{T}$ = average journey to work trip duration by car in minutes; and
P = population of urban area.

In theory, correctly specified travel demand models calibrated for one urban area should be transferable to another study area, and thereby produce traffic estimates at an acceptable level of accuracy for strategic planning purposes, but empirical research has yet to establish conclusively whether this is so. To date, only individual models in the

(a) Trip Length in Miles

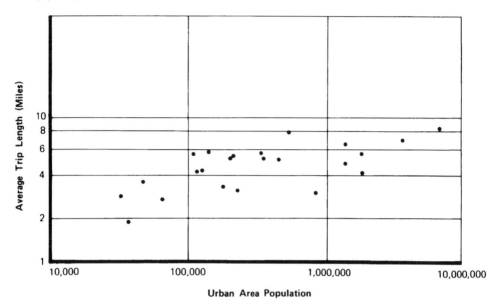

(b) Trip Duration in Minutes

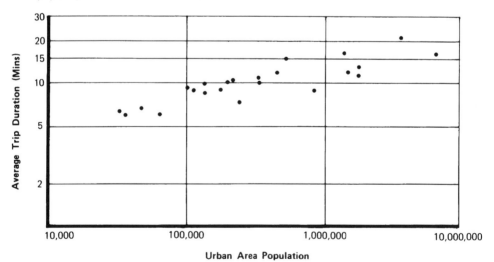

Figure 6.2: Journey-to-work Trip Lengths and City Size, North American Urban Areas
Source: Based on Voorhees *et al.*, 1967, Plate B-2, p. 89.

urban travel demand sequence have been examined; the most comprehensive coverage is by Sosslau (*et al.*, 1978). A study of residential traffic (Black and Salter, 1974) found that when zonal land-use data for Torbay were substituted into regression equations calibrated for Northampton, Worthing and Weston-Super-Mare estimates of the total number of trips by car were over-predicted by 3 per cent, 6 per cent and 7 per cent respectively.

A category analysis trip production model, calibrated with data from the 1972/3 British National Travel Survey, compared favourably with models calibrated on transport study data collected in Lincoln, Sheffield/Rotherham, Bristol and south-east Dorset (Hayfield and Stoker, 1978). Zonal work and shopping trips were estimated by the 'national' model with sufficient accuracy, but statistical tests found that the local trip rates were needed to give the equivalent accuracy for 'other' trip purposes (ibid., pp. 216-17). The sensitivity of traffic assignments to errors in zonal traffic generation was investigated by Creighton, Hamburg Planning Consultants (1971, pp. 62-4). For a doubling of the traffic at the zone centroid, the dissipation of the error on the transport network was rapid: 'the adjacent *links* average about 18 per cent more volume . . . By the time one reaches links three times removed, the differences are down to about 1 per cent' (ibid., p. 62).

When using the Bureau of Public Roads gravity model (equation (3.10)), graphs have been developed that indicate the appropriate set of 'friction factors' (Sosslau *et al.*, 1978, Figures 7-30, pp. 26-38). Traffic is stratified into home-based work, 'other' and non-home-based trip purposes, and 'friction factors' are also shown for four different-sized cities. The procedure is to measure the airline distance between two zones and look up the graph (with the correct population size) for the estimated travel time (based on differing assumptions about the proportion of traffic using arterial roads and freeways). From the estimated inter-zonal travel time, the value of the journey purpose 'friction factor' is also read directly from the graph.

One of the claimed advantages of behavioural travel demand models is 'a greater degree of transferability' (Hensher, 1977c, p. 82), but this has yet to be verified empirically. A study transferring multinomial logit models of mode choice for the journey to work, calibrated for Washington, DC, Minneapolis-St Paul and San Francisco, concluded that there was little ground to claim that the coefficients are transferable (Talvitie and Kirshner, 1978, p. 327). Binary mode-choice logit models were calibrated using pooled survey data for journey to work travel in Liverpool, Manchester, Leicester and Great Yarmouth in the hope of developing an equation of general relevance (Rogers *et al.*, 1971). The coefficients may be adjusted slightly to represent local conditions, as explained by Bruton (1975, pp. 185–9).

Simplified approaches to modelling urban travel demand have three main practical advantages. First, they eliminate data collection, especially household surveys, which avoids the cost of information acquisition. Second, they avoid the time and frustration associated with model development and calibration. Third, they allow the traffic implications of alternative land-use–transport plans to be assessed relatively quickly. A practical application of simplified procedures in the strategic land-use–transport planning process is now demonstrated.

6.3 Canberra Concept Plan Objectives

During the 1960s Canberra's population growth rate was about four times the national average: the annual average increase was nearly 11 per cent from 1961 to 1966, and nearly 9 per cent from 1966 to 1971. As noted in the previous chapter, the plan for 250,000 population was unsatisfactory, and so a better plan to accommodate growth was required (Harrison, 1964, p. 263). The probability of indefinite expansion led to the examination of the possible forms and directions of continuous urban growth. The aim was to establish a principle of growth for Canberra rather than a precise plan for the indefinite future. For the purposes of a long term study of a population horizon of one million was adopted.

The primary purpose of the *Canberra Land Use Transportation Study: General Plan Concept* was: 'to provide guidelines for the preparation for the Canberra Region of a long-range development plan which will minimise the probability of significant traffic congestion occurring as the region grows to serve populations of 500,000 and more' (Voorhees *et al.*, 1967, p. 7). A secondary goal was to provide a simple 'structural framework' to allow the city to be redeveloped if necessary to meet any social or technological changes (ibid., p. 8).

6.4 Simplified Travel Demand Models

Because of the urgency in formulating a suitable plan, no surveys were conducted, and so the traffic forecasting models contain parameters assembled from a variety of sources. Traffic generation models were based on trends observed in other cities modified where necessary by the 1961 CATS survey data (Chapter 5). Traffic distribution modelling involved taking the Bureau of Public Roads gravity models calibrated for a transport study in Washington, DC. Assumptions were made about future modal split. Traffic was allocated to links of a skeleton main road and line-haul public transport network by an *all-or-nothing* assignment.

At the traffic generation stage, four trip-purpose categories were defined—home-based work, home-based shopping, home-based 'other' and non-home-based trips. The dependent variable was the number of car-driver journeys. Logically, traffic to and from work by private transport per head of population is a function of the labour-force participation rate, the proportion of trips made by public transport and car occupancy rates. The average number of person trips to *and from* work was 0.58 per resident from the 1961 Canberra survey. 85 per cent of all person work trips were assumed to be made by private transport and car occupancy trends from six US cities resulted in the following model:

$$\hat{Q}_{pi}^{w} = 0.41 \, P_i$$

where
$\hat{Q}_{pi}^{w}$ = estimate of the daily number of home-based car-driver trips produced by zone i (both trips in and out) for trip purpose w (work); and
P_i = population of zone i.

The parameter for the model for home-based personal business and shopping trips was based on the 1961 Canberra survey. Expressed as the average number of car-driver trips per *person*, this trip rate is 0.34 but to account for an increase in car ownership, this trip rate was scaled upwards:

$$\hat{Q}_{pi}^{b} = 0.55 \, P_i$$

where
$\hat{Q}_{pi}^{b}$ = estimate of the daily number of home-based car-driver trips produced by zone i (both trips in and out) for trip purpose b (personal business and shopping).

The zonal production equation for home-based 'other' car-driver trips was derived by taking the 1961 survey trip rate of 0.31 scaling it upwards, as before, to account for increased car ownership, then scaling it downwards slightly to reflect trends in five US cities:

$$\hat{Q}_{pi}^{o} = 0.44\ P_i$$

where

$\hat{Q}_{pi}^{o}$ = estimate of the daily number of home-based car-driver trips produced by zone i (both trips in and out) for trip purpose o ('other').

The non-home-based traffic production model for the whole of Canberra was derived from data collected for ten transport studies in North American cities. The average trip rate in these cities was 1.3 trips daily per car. Multiplication of this future trip rate by a population of one million and a future car ownership level of 0.4 cars per person gave an estimate of 520,000 non-home-based vehicle trips. *Zonal* traffic production was calculated by assuming that the number of production equals the number of attractions, which were obtained from the traffic attraction model described below.

Traffic attraction modelling entailed establishing the *relative* 'attractiveness' of each zone. Home-based work traffic attractions to a workplace zone by car were based on total employment in that zone, scaled appropriately so that the metropolitan-wide total number of traffic *attractions* by car equals the metropolitan-wide total number of traffic *productions* by car.

Retail employment is one indicator of the size and hence attraction of a shopping centre. Because approximately two-thirds of shopping travel in five US cities was for convenience purchases, employment in convenience retail stores was weighted more heavily as an index of attraction than was employment in stores involving 'comparison' shopping:

$$\hat{Q}_{aj}^{b} = E_{j}^{c} + 0.62\ E_{j}^{d}$$

where

$\hat{Q}_{aj}^{b}$ = estimate of the daily number of home-based car-driver trips attracted to zone j (both trips in and out) for trip purpose b (personal business and shopping);

E_{j}^{c} = employment in zone j devoted to the retailing of convenience shopping goods; and

E_{j}^{d} = employment in zone j devoted to the retailing of comparison shopping goods.

A single convenient index of attraction for home-based 'other' trips was difficult to establish because of the complexity of travel for social, recreational and educational purposes. In three US cities 69 per cent of home-based 'other' trips were attracted to residential land uses, 16 per cent to commercial land uses, and 15 per cent to other types of land use with employment opportunities:

$$\hat{Q}^o_{aj} = 0.69\, P_j + 4.0\, E^r_j + 0.37\, E^o_j$$

where

$\hat{Q}^o_{aj}$ = estimate of the daily number of home-based car-driver trips attracted to zone j (both trips in and out) for trip purpose o ('other');

E^r_j = retail employment in zone j; and

E^o_j = all other employment in zone j.

Non-home-based trips attracted by type of land use was 42 per cent to residential land, 50 per cent to commercial land and 8 per cent to 'other' employment in two US cities:

$$\hat{Q}^n_{aj} = 0.42\, P_j + 12.3\, E^r_j + 0.2\, E^o_j$$

where

$\hat{Q}^n_{aj}$ = estimate of the daily number of car-driver trips attracted to zone j (both trips in and out) for trip purpose n (non-home-based travel).

Future inter-zonal desire-line patterns of traffic by trip purpose were estimated by the Bureau of Public Roads gravity model (equation (3.10)). Essentially, gravity models by trip purpose calibrated for Washington, DC—a city with similar government public service base to that of Canberra—were used. Friction factors for personal business and shopping, 'other' and non-home-based trips were transferred directly from the Washington transport study but for work trips adjustments were made. Journey-to-work trip lengths generally increase with city size (Figure 6.2) and for Canberra, with one million residents, the expected trip length is from 7 to 8 miles (11 to 13 km). The future average travel time by car anticipated for Canberra is from 15 to 16 minutes, assuming good roads and a low-density development. The Washington work trip friction factors were altered to produce estimates in the above travel time range.

The future inter-town public transport system was assumed to be of a very high quality, and would attract 10 per cent of work trips away from private transport, 5 per cent away from personal business and shopping trips by car, 2 per cent away from 'other' trips by car, and 5 per cent away from non-home-based trips by car. All trips combined, the future modal split on public transport was assumed as 17 per cent.

6.5 Land-use Plans

Eighteen suitable sites for 'towns' were identified by the NCDC planners. The approximate locations of these sites are shown by the diagrammatic map (Figure 6.3). Included are North and South Canberra which were already built up in 1966, two towns under construction at that time (Woden and Belconnen) and two towns across the border in New South Wales (Yass and Queanbeyan). The working assumption was that each increase of 1,000 people would require 30 hectares of residential space and 20 hectares for other uses (offices, shops, schools, industry, open space, roads and public utilities) *within* the urban area, plus a further 13 hectares outside the urban area for activities of regional importance. An additional 100 square kilometres were set aside for broad-acre uses such as airports, defence and research establishments. Overall the density within the urban area would be about 2,000 persons per square kilometre.

Six radically different metropolitan development options were prepared by the planners (Holford, 1972). The 'building blocks' for these alternatives were 'free-standing towns', each containing populations of from 50,000 to 150,000 people, which were envisaged as relatively self-supporting communities with about three-quarters of a million square feet of retail space and employment opportunities for 12,000 to 15,000 workers.

The major differences are the direction of future expansion, the number of 'towns' and the average amount of employment in each 'town' and in the central area (an area of about 400 hectares which includes the parliamentary area and government offices as well as the city centre).

There are two reasons why the NCDC planners are able to contemplate radically different urban forms with the knowledge that any preferred plan might eventuate: land is in public ownership, and government employment forms the economic base of the city. Land tenure in the Australian Capital Territory is leasehold and ordinances provide the mode of granting leases, rights and obligations of lessees,

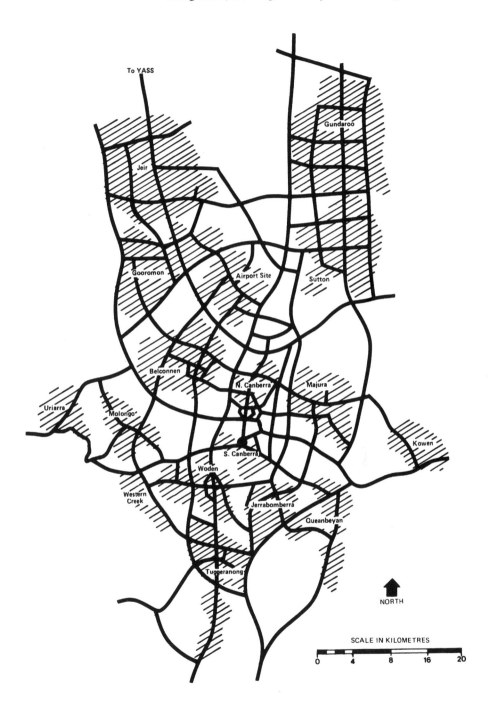

Figure 6.3: Land-use and Transport Options, Canberra, at the One Million Population Level

building controls and the like (Brennan, 1971). Nearly 60 per cent of the workforce are public servants or are on the government payroll (Harrison, 1978, p. 110). Although the growth in public service employment is determined by government policy, the NCDC has some power of persuasion in getting government departments to establish offices in suitable locations.

6.6 Evaluation of Land-use Alternatives

The six development options were assessed primarily on their traffic implications. Traffic forecasts were made for each land-use alternative using simplified travel demand models but the differences among the six alternatives are small: Plan A (concentric development) produces the minimum amount of daily travel −15.6 million vehicle-kilometres− but this is only 2 per cent less than the maximum figure which is generated by Plan A-2 (centralised employment). Estimated average travel times by car are also within a very narrow range of from 10.8 to 11.7 minutes. For the journey to work by car, corridor growth would involve about 15 per cent more travel time than a concentric development. Average daily travel costs varied by as little as 5 per cent.

Because the analysis did not show clear differences, five questions were asked. What are the general transport consequences of alternative urban forms? What are the transport implications of different employment levels in the central area? What are the transport implications of different employment levels in town centres? Where do traffic 'bottlenecks' occur? What is the potential for public transport under alternative urban forms?

Answers to these questions helped in the formulation of an urban development strategy. In terms of the arrangement of 'towns', the corridor concept was preferred because a linear form allows the maximum potential of public transport to be exploited, and allows freeways to be located on either side of the urban corridor, thereby reducing their adverse effects on residential neighbourhoods.

Central area employment was not of crucial importance when trying to identify potential traffic problems in a city of one million. Traffic generated by a central area employment of 120,000 would be within the capability of the proposed roads converging on the centre. However, this judgement was based on average daily traffic assignments and clearly peak-hour difficulties would emerge well before this build-up of employment was reached.

Towns are not very self-contained when their populations are below

75,000 because of out-commuting; when they become too big, roads leading to the town centre become congested with traffic. The optimum range of population for 'towns' appears to be from 100,000 to 120,000 residents, with 10,000 to 15,000 job opportunities.

Transport bottlenecks were found to occur primarily around town centres if they were located next to a central transport spine which functioned both as an arterial for inter-town traffic and as a collector of local residential traffic. Much non-essential traffic would be concentrated at the town centre, because of the few intersections through which the majority of traffic could pass, irrespective of its ultimate origin or destination.

Public transport was expected to play an important part in the final transport system. Concentric urban forms require radial public transport routes, and these 'branch lines' are a barrier to an efficient design because the average daily loading of passengers is less than 15,000 per mile. For this reason, a linear arrangement of 'towns' has advantages in allowing higher levels of service on public transport.

This critical appraisal led to a set of principles being established to guide the formulation of a more suitable urban development strategy.

(a) A corridor arrangement offers advantages in exploiting a highly structured transport system.
(b) The central area could accommodate an employment level of 120,000, but a level of 90,000 is preferred, thereby avoiding peak-period congestion.
(c) The recommended size for individual towns is from 100,000 to 120,000 residents with 10,000 to 15,000 jobs.
(d) A road system established on the hierarchy concept of major highways to channel the longer journeys from one town to another, of arterial roads for traffic within a town, and of local access roads to serve residential areas and other land-use activities.
(e) In order to exploit the potential of public transport to its fullest: future urban growth should be directed into corridors to increase the density of point-to-point travel demands; as much employment and other activities as possible should be concentrated in the central area, because the city centre is usually the single most accessible place reached by public transport; and 'town' centres, other non-central employment areas and important traffic generators should be located along the major public transport spine.

The major concept to emerge (Plan F) consisted of three corridor groupings of towns radiating in a Y-form from the central area, plus the older, partially disconnected town of Queanbeyan. A slight variation of this plan was also prepared (Plan G) which allowed for the possibility of a new airport at Gungahlin. These two concepts were evaluated in an identical manner to the other six concepts except that the road network was modified slightly to distinguish a road hierarchy comprised of freeways and arterials. Table 6.1 summarises the transport system performance of Plan F and Plan G together with Plan A which has been included for comparative purposes. For Plan F, the average travel time for all trip purposes was lower than the average for five of the six previous options and total vehicle-kilometres were also lower. Average daily travel costs are higher than Plan A because the freeways entail higher unit construction costs.

Table 6.1: Measures of Transport Performance–Canberra General Concept Plan and Alternative Options

Daily Transport Performance	Alternative Urban Development Concept		
	F	G	A
Average Travel Time (Minutes)			
Home-based work	16.0	16.2	15.8
Home-based shopping	8.4	8.4	8.5
Home-based other	11.3	11.3	11.2
Non-home-based	10.2	9.7	10.9
All trip purposes	11.2	11.2	11.3
Vehicle-km (thousands)	15,310	15,460	15,600
Airline-km (thousands)	10,650	10,140	10,700
Travel costs (thousands)	$670	$680	$657

Source: Based on Voorhees *et al.*, 1967, Table 13, p. 60.

Plan G was not an attractive proposition for two main reasons. First, the relocation of residents and employment from Gungahlin (airport site) further out to the north-east at Gundaroo required more high capacity roads at extra cost. Second, the potential patronage on public transport was insufficient to justify the costly extension of the system on its own right of way out to Gundaroo.

The arguments for and against all of the urban development options are set out in *Tomorrow's Canberra* (NCDC, 1970), but the preferred option (and the one which is still guiding long-term development) for the general concept plan is the Y-Plan. Its transport principle is a public transport (rapid transit) spine linking town centres, 154 km of freeways located mainly outside the urban corridors, and urban arterial roads within each 'town'.

Inter-town freeways were preferred to expressways because an analysis of construction, maintenance and vehicle operating costs of urban arterial roads, expressways and freeways for a wide range of traffic volumes indicated that only for traffic volumes up to about 20,000 vehicles per day were four-lane urban arterials a more economic solution than freeways. As shown by the traffic assignments, major links were expected to carry from 35,000 to 40,000 vehicles per day, with a few carrying upwards of 70,000 vehicles per day.

6.7 Implementation

The NCDC adopted the Y-Plan concept as a guide for the long-term development of Canberra. An important aspect of the analysis contained in the *Canberra Land Use Transportation Study* was the staging of future growth, and the recommendation that 'towns' be developed sequentially, although with some overlap. This led to an intermediate plan for half a million people, consistent with the framework established by the general concept plan, and the best staging of development.

In 1969, a decision had to be reached about the next 'new town' to be built because in Woden the servicing of blocks for low-density housing was complete, and in Belconnen, the second 'new town', there was every indication that residential servicing would be completed by the mid-1970s. The two options for the next stage were Tuggeranong in the south and Gungahlin in the north. There were advantages and disadvantages on both sides but Tuggeranong was chosen primarily because the latter site was under consideration for a future airport—a proposal which was subsequently abandoned.

Figure 6.4 shows the developed areas of Canberra and Queanbeyan to June 1979, when the population was an estimated 230,000. From mid-1958, when the NCDC came into full operation, the construction expenditure in the following 21 years mounted to about $A2,230 million (1979 dollars). The spatial expression of this expenditure shown in Figure 6.4 are the 'new towns' of Woden-Weston Creek, Belconnen and Tuggeranong (under construction). The three major 'town' centres already built are City Centre, with $68,500 \, m^2$ of retail space, Woden Town Centre with $71,500 \, m^2$ of retail space, and Belconnen Town Centre, which, with the opening of Stage Two of Belconnen Mall early in 1979, has about $65,000 \, m^2$ of retail space. Some progress towards the proposed inter-town public transport system is apparent with the

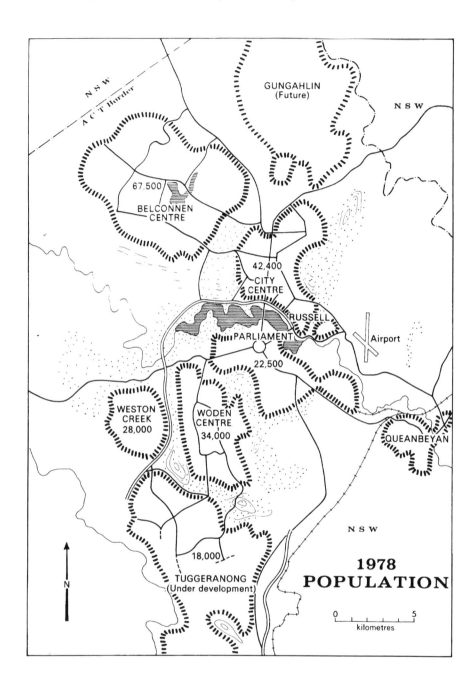

Figure 6.4: Canberra, as Developed

construction of the Woden and Belconnen bus interchanges, the separate busway through Belconnen Town Centre and the bus-only lane connecting Woden Town Centre with the city centre. The completed part of the peripheral freeway plan is the Tuggeranong Parkway and the Molonglo 'Arterial' on the north side of Lake Burley Griffin (Figure 6.4).

Future population growth will be accommodated first in Tuggeranong and later in Gungahlin, but beyond the half-a-million mark, the Y-Plan extends north-east and north-west over the border into New South Wales. The federal government announced in March 1976 that all ideas of extending the Australian Capital Territory would be abandoned. 'Without public acquisition of all or most of the areas needed for urban expansion—and their rural surrounds—the Y-Plan or any other purpose- ful planning arrangement across the border is barely worth pursuing' (Harrison, 1978, p. 117). Consequences of cramming a population of one million into the Australian Capital Territory were contemplated by the NCDC in *Metropolitan Structure Plan Review* (NCDC, 1976a).

Whilst the findings have not altered the Commission's policy towards long-term development, the interpretation of technical analyses sheds light on to the role of mathematical models in the planning process. The Y-Plan and a compact development pattern were evaluated in a similar way to that described in section 6.6, except that traffic implications estimates were by TRANSTEP—a land-use—transport interaction model (Nairn *et al.*, 1978)—with up-to-date travel parameters for Canberra. More sophisticated modelling gave results that were identical to the *Canberra Land Use Transportation Study* but the conclusion drawn was different: Canberra's growth should be contained entirely within the Australian Capital Territory—an expedient recommendation.

6.8 Summary

Strategic transport planning examines long-term land-use and transport alternatives in a broad-brush way, with the analysis confined primarily to the spatial pattern of homes and workplaces and the major transport system, to determine the best urban development strategy. Because the future is uncertain it is important to consider and evaluate different strategies. The models outlined in this chapter, which are deliberately more simplified than the ones described in Part One, provide an efficient way to analyse a wide range of urban development concepts.

This type of transport planning is more relevant in the planning of new towns or where very rapid urbanisation is occurring. A case study, the *Canberra Land Use Transportation Study*, has been presented. The Y-Plan is a flexible framework and has been accepted by the NCDC as a guideline to direct urban growth in a purposeful way. Long-term plans for public transport are discussed in the next chapter. Transport planning studies with shorter time-horizons are described in Chapters 8 and 9.

7 PLANNING FOR PUBLIC TRANSPORT

The main topics covered in this chapter are demand forecasting methods, the evaluation of public transport technologies and the use of attitudinal surveys in planning for public transport. These correspond to very broad levels of planning for land use and public transport, and to the functional planning of route location and service levels. More detailed operational planning, such as vehicle scheduling, duty rostering, timetabling and marketing, is not discussed here (see Shortreed, 1974, Chapters 5, 6 and 10).

The first section outlines historical trends in patronage and operating costs and indicates the difficulties facing public transport operators in planning for the future. The conventional wisdom is that better public transport services will attract travellers, and the examples of public transport planning in Canberra reflect this optimism.

7.1 Urban Public Transport Prospects

During the post-war period, there has been a marked decline in public transport patronage, even though population has increased. The annual number of revenue passengers carried by surface public transport in the USA was at a peak of 18,982 million in 1945 but dropped by 69 per cent to 5,932 million in 1970. In Canada, the peak of 1,290 million passengers was reached in 1950, but this had fallen to 1,000 million by 1970 (Garner, 1974, pp. 7-9). Municipal operators in Britain carried 7,000 million passengers in 1950-1 and 4,200 million in 1968-9 (White, 1976, p. 34).

Another indicator of public transport patronage is the number of trips *per capita*. In 17 major cities throughout the world the annual number of trips *per capita* has been falling consistently, except in Tokyo and Hong Kong (Thomson, 1977, Figure 7, p. 86). In the five mainland state capitals of Australia, the use of public transport has fallen from 327 annual trips per head in 1954 to 152 annual trips per head in 1971 (BTE, 1972, Annex A). The corresponding figures for the same period in US urban areas were 114 and 38 (Quinby, 1976, Figure 6.9, p. 249). In 60 British towns, the decline in the number of bus trips *per capita* in the decade from 1960-1 was about 30 per cent, although some towns with better public transport marketing policies had halted this downward trend by the 1970s (White, 1976, p. 37).

Whilst patronage has been falling, public transport operating costs have been soaring, primarily because the industry is labour-intensive. For example, in major Australian cities the operating costs of bus undertakings have trebled during the last decade. In Melbourne the average operating cost per bus-kilometre for government services rose from 43 cents in 1970-1 to 111 cents in 1977-8; for private companies (who operate about 30 per cent of the buses), the increase was from 26 cents in 1970-1 to 55 cents in 1976-7 (Wallis, 1979, Figure 1, p. 709). Lower costs of private operators are attributed to greater flexibility and efficiency in the deployment of labour, and lower award wages, pension and retirement payments.

The widening gap between public transport operating costs and revenues has required massive government subsidies to maintain services to a current standard. In Australia, suburban rail, bus and tram services operated by government authorities are losing money (although private bus companies apparently made a profit), with the government contributing 41 per cent of costs in 1970-1 (Clark, 1975, Table I, p. 453). Although more up-to-date figures are unavailable, every indication is that this proportion has risen. For example, the cost recovery with Melbourne's trams is 55 per cent and with Sydney's buses, 36 per cent (Caldwell, 1979, p. 699).

Various social and economic factors have added to the difficulties facing public transport operators. In the post-war period the advantages of private transport—flexibility, privacy and low labour costs—have resulted in increasing car ownership and use, despite its higher resource cost per passenger kilometre. In addition, public transport fares have increased relative to general price rises (Thomson, 1977, p. 87) and to motoring costs (Neutze, 1977, p. 136). Changing social patterns have eroded the potential demand for public transport: a six-day week has been replaced by the five-day work week; shorter lunch breaks make it difficult for workers to go home at midday, and virtually impracticable as cities grow bigger; and attendance has fallen at sporting events and cinemas.

Low-density residential suburbs and the suburbanisation of jobs and services are creating cities that are becoming increasingly unfavourable to the efficient operation of public transport. Railways operate profitably and work best in cities with high population densities and activities concentrated in the CBD. Tram routes also require a high traffic density to justify the building and maintenance of tracks and power sources.

Buses are more flexible to serve the low-volume, diffuse movement patterns in suburban areas.

Public transport operators have always been faced with catering for temporal fluctuations in demand, but the loss of off-peak patronage has enlarged the peak demand for public transport. In addition, there is significant peaking within the rush hour itself, and so extra capacity must be provided. Also, there is usually an unbalanced directional flow on routes inwards in the morning and outwards from the city centre in the evening, resulting in empty seats. Capital equipment is underutilised because some of the vehicles required to meet peak demands remain idle at other times of the day. For example, in Sydney, 1,380 government buses are in use at 5 p.m., but by 9 p.m. the number has fallen to 280 (Webb, 1975, p. 7). Labour is underutilised because the peak extends over six or fewer hours and two peaks are sufficiently apart to require split shifts, which increases labour costs (with penalty wage rates) and the amount of unproductive time (staff signing on and off and vehicles travelling to and from depots without passengers).

Notwithstanding these indicators of future prospects for urban public transport, some planners have argued enthusiastically that system extensions, improved service frequencies and new technologies will attract people back to public transport. Whether this is based on personal preference for public transport, guilty feelings for ignoring public transport planning in the past or a genuine belief in forecasts from modal-split models is largely immaterial: the fact is that transport planners have largely avoided the difficult issues of defining the social service obligations of public transport or of reducing the deficit. The studies described in the remainder of this chapter are examples of the dominant planning philosophy of the 1970s—better public transport.

7.2 Passenger Forecasting in Canberra

The responsibility for planning public transport in Canberra is divided between the NCDC and the Department of the Capital Territory (formerly the Department of the Interior). In 1972, a Transport Policy and Planning Branch was established within the DCT and the planning roles of the two organisations became more formally co-ordinated, with the NCDC 'concerned with the panorama and the DCT with specifics' (Webb and Cooper, 1976, p. 7).

Bus routes extend to all suburbs. Feeder services on suburban routes generally link with the town centres or bus interchanges, and inter-town routes connect the town centres as directly as possible on the arterial road system. Peak headways are from 15 to 30 minutes (7 to 8 minutes on inter-town services) and off-peak headways are about 30 minutes (15 minutes on inter-town services). Despite the fact that about two-thirds of Canberra's population are at least occasional users of public transport (Wardrop, 1979) school and commuter services required a 25 per cent subsidy, and weekend and off-peak services required a 75 to 80 per cent subsidy for 1976-7.

At present, buses operate in Canberra, but as noted in the previous chapter, some form of line-haul public transport is envisaged. More detailed information on likely patronage levels on a new system was required. The *Canberra Public Transport Study* (1974) provides a good example of forecasting the long-term demand for public transport, and the way that functional characteristics of future transport systems are modelled. The key objectives of this study were to determine the appropriate technology to operate on the right of way reserved for inter-town transport as part of the Y-Plan (Chapter 6), and to assess the resources required to implement any new system.

The traffic forecasts were based on the working assumption that Canberra reaches a population of one million by the year 2010. Within the framework established by the Y-Plan, the NCDC planners allocated population, residential work-force and 450,000 jobs (including two alternatives of 70,000 and 100,000 jobs to the central area) to 344 traffic zones. The traffic estimation model comprised a zonal traffic generation model, a Bureau of Public Roads traffic distribution model, a modal-split diversion curve and a capacity restraint traffic assignment model.

The daily zonal traffic generation model was specified as follows. Home-based work trips produced were 1.8 per worker based on the assumption that (a) every worker made two work trips each day; (b) 5 per cent made additional trips by going home for lunch; (c) absenteeism accounted for 7 per cent of the work-force; and (d) 8 per cent walked or cycled to work. All other home-based trips and non-home-based trips were assumed to equal 1.1 trips per *person*, based on trip rates in US cities. The zonal traffic attraction model for work trips was equal to the number of jobs in the destination zone. The attraction index for all other trips was one-third of the zonal residential population plus two-thirds of the zonal employment.

Trip-purpose friction factors used in the Bureau of Public Roads gravity model (equation 3.10)) were transferred from the *Perth Regional Transport Study 1970* (see section 6.2). The modal-split diversion curve (see section 3.4.1) was also transferred from another transport study conducted in Los Angeles. The consultants looked towards an existing city with very high car ownership, and selected Los Angeles, which had 0.43 cars per person. The general shape of the Los Angeles modal-split diversion curve was followed, although the proportion using public transport was increased very slightly to reflect a land-use pattern in Canberra that was slightly more supportive to public transport. Person trips were converted into private vehicle trips by assuming future car occupancies.

Generalised cost in 'equivalent' minutes (see section 2.3.2) was used in the traffic distribution, modal-split and traffic assignment models. Public transport fares, motoring costs and parking charges were converted into 'equivalent' minutes by valuing travel time at 25 per cent of the wage rate. Modal split was estimated from the difference in 'equivalent' minutes by car and by public transport. The important feature of the traffic assignment model was the inclusion of traffic flow-dependent travel times by road, and different levels of service by public transport. Initial headways were specified on the public transport network, but these were adjusted in response to the peak-load requirements on each link. Thus congestion effects on the roads and different frequencies of service on public transport influenced inter-zonal travel times, which in turn altered the traffic pattern via the feedback mechanism of travel times in the traffic distribution and modal-split models (see section 4.4).

Traffic forecasts were made assuming three different conditions of transport supply. The road network was identical in all three cases, but the operational characteristics of public transport were varied to reflect the provision of buses operating on surface streets, or buses operating on grade-separated structures, or a fully automated intertown system. These alternatives were evaluated by comparing likely passenger revenue against construction and operating costs.

Capital costs were about \$A20 million for the surface street system, \$A50 million for busways and \$A79 million for the automated system (\$A90 million on elevated structures). Annual operating costs were expected to be \$A10.6 million, \$A7.7 million and \$A3.4 million respectively. The net present value (equation (4.21)) of each alternative was calculated for both a low modal split of 13 per cent on public transport and a high modal split of 40 per cent and for discount rates of 6 per cent and 10 per cent per annum. For all assumptions, the net present value was found to be negative. Buses at grade would be clearly the least uneconomic investment for low patronage, and automated systems would be justified only for high patronage and a low discount rate. Nevertheless, the *Report by the Steering Committee* favoured the automated system (Voorhees *et al.,* 1974a, p. 7) but suggested further investigations to assess more accurately the commercial availability and costs of automated public transport systems.

7.3 Evaluation of Alternative Technologies

Although dreamers, artists and planners have shared common visions of futuristic public transport technology, a serious evaluation of the feasible systems is essential to achieve the best use of resources. The Canberra *Intertown Public Transport* (IPT, 1976) was a timely report because there was an urgent need 'to resolve the issue of public transport technology to ensure that imminent land use developments ... and bus infrastructure provision ... would be compatible with the eventual form of any upgraded public transport system' and to reformulate transport policy because of 'the social and environmental pressures and feedback regarding the energy implications of a low density/low public transport use city' (ibid., p. 17).

A major difficulty encountered was the identification of suitable technologies. Although details of conventional buses and railways (vehicle size, performance, track and signalling requirements, capital and operating costs, etc.) were easy to come by, information about 'non-conventional' automatic systems was more difficult to obtain from the available technical literature. Consequently, questionnaires were sent to 51 known manufacturers of 'non-conventional' transport equipment, requesting information about design concepts and operational performance.

The information obtained allowed 21 'non-conventional' systems to be screened in terms of three fundamental functional criteria: whether they were capable of a minimum cruise speed of 50 kph; whether they could handle the forecast peak-hour patronage levels expected in Canberra towards the end of this century; and whether they could be installed and made operational no later than the mid-1980s. The First Assistant Commissioner for Engineering, Dr O'Flaherty, said emphatically, 'non-conventional systems ... were either too inefficient or not developed yet. There was nothing to be gained using Canberra as a proving ground' (*Canberra Times*, 27 July 1976). Because 'non-conventional' technologies were ruled out, more conventional forms of public transport were considered: bus priority, busways (with either standard or articulated buses), railways and tramways (either at-grade or grade-separated).

A range of likely patronage levels was assumed for all systems. Low and high public transport modal splits were calculated from assuming radically different urban development concepts and transport policies. Traffic forecasts are an important input when assessing the benefits to travellers of transport improvements. The economic evaluation involved comparing the time-stream of benefits and costs of the four alternative public transport systems (see section 4.6). The 'do-nothing' case was that future public transport services would be provided by buses competing for road space with motor cars and lorries.

In practice, the identification and quantification of all relevant costs and benefits is a tricky business. The *Intertown Public Transport* study calculated capital costs from the following:

Civil Works costs for the alternative system considered *minus* equivalent costs for the bus-in-mixed traffic base case,
plus Costs of vehicle storage and maintenance facilities for the alternative system *minus* equivalent costs for the base case,
plus Costs of initial and additional fleet vehicles for the alternative system *minus* equivalent costs for the base case,
plus Capital cost of complementary buses required for the alternative system during construction,
less Adjustment to previously calculated fleet costs to reflect pull-on capability (applicable to bus systems only) (IPT, 1976, pp. 71–2).

Capital costs were for all engineering works on a provisional route alignment, including earthworks, structures, track or pavement, stations, service alterations, drainage, property acquisition, electrification and signalling work, all based on local unit-cost rates. Costs of vehicle storage and maintenance facilities for buses were estimated from the construction costs of the Belconnen bus interchange. Fleet costs were obtained from the initial present value of the number of vehicles required each year. 'Complementary buses' refer to the capital costs of providing a back-up service during the construction period (from 1983 to 1990) of a railway or tramway. An advantage of all bus systems is that the same vehicles can perform both feeder and line-haul services and corrections were made to reflect a smaller vehicle fleet.

Because costs are restricted to capital expenditure, any other 'costs' were called 'negative benefits' and were subtracted from the following list of benefits, as indicated:

Savings in operating costs due to the alternative system considered, i.e. total present value of the operating costs of the bus-in-mixed traffic base case *minus* equivalent costs for the project system,

plus Benefits to passengers due to reductions in travel time,

plus Benefits to passengers converting from private to public transport because of the development of the alternative system,

plus Benefits to generated passengers (in this study assumed negligible),

plus Savings in replacement cost of vehicles,

less Disbenefit due to increased station manning and maintenance costs of the alternative system,

less Disbenefits due to interchanging penalties during construction (rail systems only),

less Disbenefits due to lack of pull-on capability (rail systems only),

less Disbenefits due to increased passenger waiting times because of lesser frequency of service (rail system only),

plus Corridor benefits i.e. benefits which the system brings to travellers other than public transport users who travel in the urban corridor through which the line-haul system passes (e.g. other road users),

plus Benefits due to the residual value of the project (infrastructure, vehicles, equipment and operations) (IPT, 1976, p. 72).

The derivation of the seven positive benefits are described in order.

First, a major benefit of better transport is the reduction in the cost of real resources to run a new system compared with the resources consumed in operating buses under mixed-traffic conditions. Unit bus operating costs were supplied by the Department of the Capital Territory, and increased by one cent per kilometre to allow for extra road maintenance caused by the wear and tear of buses. Railway unit operating costs were based on the New South Wales Public Transport Commission figures, except they were adjusted downwards by 30 per cent to reflect probable one-man train operations in Canberra. Unit operating costs of trams were 'interpreted' from current tramway operations in Melbourne and in Sydney (which had trams until 1961). Annual operating cots for these systems were obtained by multiplying unit costs by the estimated vehicle-kilometres travelled each year.

Second, another major benefit is savings to public transport users. The difference in average travel time from town centre to town centre for buses in mixed traffic, assuming an average running speed of 35 kph and the new system, all assumed to perform identically at 50 kph, was multiplied by the number of passengers forecast on each section to give total travel time savings in minutes. A monetary value of travellers' time, taken as 1.5 cents per minute (16 per cent of the 1975 average hourly male wage rate), was used to convert time savings into cost savings.

The third benefit listed refers to travellers who previously made their journey by another transport mode but diverted to inter-town public transport. However, only the grade-separated systems were expected to attract passengers away from private transport and the benefits were small. At most, savings in travel time for these 'converted' passengers amounted to a discounted $A1.4 million for articulated buses operating on a busway.

The fourth benefit listed is the number of 'induced' or 'generated' trips, but these were assumed to be negligible. A fifth benefit, savings in vehicle replacement costs, is included because railway and tramway carriages have an expected life of 30 years; buses need replacement after 12 years.

The sixth benefit listed accrues to users of private transport who live close to any grade-separated inter-town public transport system because of the removal of buses from the roads. Traffic flows were expressed in passenger-car-equivalents, with and without buses, and the difference in travel times and vehicle operating costs were multiplied by the number of vehicles to give an estimate of total benefits. For a low usage of public transport these non-user benefits to the year 2000, discounted at 7 per cent, would be $A1,137 million; for a high usage of public transport the total benefits would be $A2,921 million (ibid., p. 77).

The final benefit in the list is the salvage value of the system. Allowance was made for the residual value of the total project at the end of the formal evaluation period: as fixed capital works have no salvage value, benefits calculated for the year 2000 were continued at the same level for an additional 30 years.

Also included in the list of benefits are four items identified as 'negative benefits', and their derivations are now briefly described. One, station manpower costs were equal for all systems at $A86,000 per 'town' station per annum, and maintenance costs were varied for each system, being based on an annual charge of 2 per cent of the capital cost of the station. Two, inconvenience to travellers during construction of rail systems was represented by an inter-modal transfer 'penalty' of 10 cents per passenger—based on a transfer time of three minutes and 'weighted' by a factor of two. Three, as railways cannot use feeder buses on their track, an inter-modal transfer penalty of 10 cents per passenger was included. Four, railway and tramway signalling and safety requirements necessitate greater headways than buses, and the extra waiting time of four minutes, 'weighted' by a factor of one-and-a-half, was costed as 6 cents per passenger.

The results of drawing together all the costs and benefits of the public transport alternatives are summarised in Table 7.1. The first year of expenditure was assumed to be 1979, and the costs and benefits were discounted back to that year by 7 per cent per annum—the standard rate stipulated by the Australian Bureau of Transport Economics. Passenger forecasts were made for a population of 797,000 by the year 2000, and for both high and low public transport modal splits. With low patronage, bus priority is the best system with a benefit/cost ratio of 1.5 and the at-grade tramway is the least desirable with discounted costs far outweighing discounted benefits. With high patronage, all systems have benefit/cost ratios greater than unity, but bus priority again has the highest ratio. On the other hand, articulated buses operating on a busway produce the largest positive difference between benefits and costs.

Table 7.1: Discounted Costs and Benefits of Inter-town Public Transport Systems, Canberra, 1979–2000

Public Transport Alternative	Low Patronage Forecasts			High Patronage Forecasts		
	Benefits ($A in millions)	Costs	B/C ratio	Benefits ($A in millions)	Costs	B/C ratio
Bus priority	26.8	17.8	1.5	48.2	15.6	3.1
Busway:						
standard bus	42.9	42.0	1.0	73.4	39.2	1.9
articulated bus	55.9	43.8	1.3	97.4	42.0	2.3
Railway	50.6	55.0	0.9	94.4	64.4	1.5
Tramway:						
at-grade	22.7	40.5	0.6	47.7	49.0	1.0
grade-separated	37.2	52.3	0.7	67.2	59.5	1.1

Source: Based on IPT, 1976, Table 12.1, p. 86.

Sensitivity analyses were conducted because any economic evaluation suffers from the uncertainty surrounding the future numerical values of the various factors under consideration. Population forecasts were varied from 650,000 to 900,000, capital costs of the rail systems were reduced by 20 per cent and operating costs of the rail systems were reduced also by 20 per cent whilst bus operating costs were increased from 7 to 15 per cent. Three different monetary values of time were examined. Variations in population, modal split, capital costs, the residual value and the 10 per cent discount rate all had a large effect on the absolute values of benefit/cost ratios, but did not change the ranking of the public transport systems. The ranking of projects was slightly more sensitive to changes in the value of travel time and vehicle operating costs, but bus priority was consistently ranked first, except when reduced railway operating costs of 35 per cent and high patronage resulted in that system having the highest benefit/cost ratio.

Environmental considerations were also included in the appraisal of alternative transport systems. The technique used, called the 'weighted value method', lists the environmental impacts of each plan, such as noise, pollution, visual intrusion and so on, and assigns a score out of an arbitrary maximum to reflect the planners' subjective judgement of the relative 'importance' of each impact. The higher the number assigned, the greater the environmental nuisance. Buses in mixed traffic conditions would be the least satisfactory because of road congestion, noise and air pollution; bus priority and busways largely avoid these problems. Railways and tramways would be the most satisfactory systems from an environmental point of view.

Taking all factors into consideration, priority buses emerged as the recommended system. It was clearly the best from the economic evaluation. On environmental grounds, there were significant differences between the systems: grade-separated systems were better in the long term. A grade-separated system, especially a railway, offered the operator advantages in safety, reliability, scheduling and management:

> a conventional priority bus system is the preferred system for the development in Canberra of an intertown public transport system until passenger volumes in the range of 7,000-8,000 passengers per hour are foreseen within about five years . . . The uncertainties associated with population projections and land development mean that the priority improvements should be made first at locations where they are most required and can be justified in respect of a range of likely developments (IPT, 1976, p. 118).

Bus priority measures are described in the next chapter.

7.4 Attitudinal Surveys

Attitudinal studies provide an extensive source of data both on the relative importance of transport attributes as perceived by the traveller, and on residents' preferences for alternative transport policies. Although the reliability of attitudinal measures is often questioned, this type of study can help public transport operators to identify deficiencies in their services, or can involve the public in the transport planning process by establishing their attitudes towards different land-use and transport policies.

An attitudinal survey was conducted in 1966 at dwellings in Canberra. The 1,760 usable questionnaires, a 60 per cent response rate, represented the reactions to prepared statements about bus services by a 3 per cent sample of residents aged fifteen and older. The questionnaire contained 27 statements and each respondent was asked to tick one of five squares (labelled 'strongly agree', 'agree', 'not decided', 'disagree' and 'strongly disagree') which most closely describes the respondent's feelings regarding the statement.

Table 7.2: Attitudes to Prepared Statements about Public Transport, Canberra, 1966

Prepared Statements about Public Transport	Rating[a]
Missing the last bus in the peak period results in having to wait too long for the next bus.	4.1
With increasing traffic congestion it will be more pleasant to travel on buses and not worry about the problems of driving.	3.7
It is reasonable to expect passengers to walk five minutes to a bus stop.	3.7
Bus routes wander about unnecessarily making journey times too long.	3.6
Express buses would make bus travel much more convenient for me.	3.6
Having to stand deters me from using buses.	3.0
There are enough buses during peak periods.	2.8
The provision of air conditioning in buses would be completely unnecessary.	2.5
There are enough buses during off-peak periods.	2.2
If I travel to work by bus my acquaintances may think I cannot afford a car.	1.9

Note: a. 5 = 'strongly agree'; 4 = 'agree'; 3 = 'undecided'; 2 = 'disagree' and 1 = 'strongly disagree'.

Source: Based on Pak-Poy *et al.*, 1967, Table 1, p. 7.

Table 7.2 reproduces some of these statements and indicates the average rating for each. A response of 'strongly agree' was given the rating 5, a response of 'agree' 4, and so on. Although there was a slight difference of opinion between regular and infrequent users of public transport, the general consensus was that public transport could be improved by: (a) a better spread of buses, and more buses in the peak periods; (b) more frequent bus services in off-peak periods; (c) introduction of express bus services; (d) more direct bus routes; (e) provision of air-conditioning in the buses; (f) more publicity of timetables; (g) provision of more bus stop shelters; (h) reduction in delays in boarding and paying fares; and (i) more comfortable seating. These deficiencies assisted the consultants to prepare an action plan for public transport, which to a large degree has been implemented.

Attitudes and support for various land-use and transport policies were investigated by the *Canberra Modal Split Study* (Paterson *et al.*, 1974) using a game specially designed to reveal residents' preferences for alternative policies. One side of a board showed drawings to represent schematically seven transport alternatives; the other side showed drawings to represent combinations of three housing densities and three levels of transport investment. Each person interviewed was given counters worth $A500 and asked to allocate them amongst the various options. The rules stipulated the maximum allocation to each alternative, that at least $A150 must be allocated to roads, and that not all of the $A500 need be allocated to transport. This is not really a game in the true sense because there is no competition element, but it is a simulation game because respondents are simulating the planner's decision-making procedures.

The game was played by 673 residents, who allocated counters to the seven transport options as follows:

	Roads	Train/Bus	Bus	Dial-a-bus	Parking	Paths	Cycle Paths
Maximum	$A400	$A290	$A270	$A100	$A 60	$A 30	$A 30
Average	$A235	$A 66	$A 53	$A 22	$A 31	$A 17	$A 7

Interpretation is difficult, whilst the largest single allocation is for roads, the ratio of the amount allocated to the maximum possible amount is highest for footpaths (0.57) and parking (0.52). The consultants concluded that residents were extremely reluctant to allow future road standards to fall because the most popular solution was improved roads.

The results of residents' preferences for different combinations of residential density and transport (roads and public transport) quality showed that three options were of almost equal popularity: low density with medium transport quality; low density with high transport quality; and medium density with a high transport quality. Altogether, 91 per cent of residents wanted better transport and 44 per cent wanted higher residential densities. Seven out of ten students preferred medium-density housing and both young and old alike expressed a desire for something different to the present low-density development.

7.5 Summary

Trends in public transport patronage and operating costs are important historical reminders to haunt transport planners when calculating future revenues and costs. Passenger forecasts are one aspect of planning for public transport. Attitudinal surveys provide information to public transport operators on ways of making more immediate improvements to public transport.

The Canberra *Intertown Public Transport* study is a good example of the methodology of the economic evaluation of transport because the costs and benefits of each alternative were explained in more detail than was possible in Chapter 4. The environmental and social evaluation is a far less satisfactory case study, not least because of the highly subjective, and perhaps arbitrary, scores assigned to environmental nuisances. Further examples of the economic evaluation of bus priority measures follow in the next chapter.

8 SHORT-TERM TRANSPORT PLANNING

The name 'short-term transport planning' suggests activities concerned with the geometric design and implementation of transport projects that form integral parts of a much longer-term plan, and this is the precise meaning of short-range planning as defined by the US Department of Transportation's 'Urban Transportation Planning System' (Dial, 1976, pp. 61–2). However, short-term transport planning also means planning activities aimed at making a better and more efficient use of *existing* transport facilities, or at recommending capital works that are relatively inexpensive. This is sometimes called transport system management (TSM) or urban traffic management.

This chapter elaborates on the meaning of short-term transport planning and describes various measures that can be implemented. As discussed later, the measures are wide-ranging, but three Canberra case studies are included to illustrate this type of planning. They are the planning of an area traffic control scheme, bus priority measures and the *Canberra Short Term Transport Planning Study*, which is an example of the application of behavioural travel-demand models in transport policy analysis.

8.1 Short-term Planning Studies

Short-term transport planning is concerned with the best use of existing urban transport facilities and how to improve their performance. It is also concerned with new capital works and improvements of a limited scale which are relatively inexpensive and are capable of implementation immediately, or at least within a couple of years.

The longer-term transport planning studies described in the preceding three chapters are concerned primarily with large, capital-intensive proposals that, if implemented, would inevitably entail several years of preparatory work in planning, design, project management and construction. As noted by Wells (1975, p. 42), 'there is often a hiatus in transport planning implementation at the start of a plan period. This gap can well be filled by an Immediate Action Programme ... comprising relatively inexpensive traffic management measures.' Short-term programmes, designed for up to five years, are envisaged as having significant and immediate impacts.

181

Short-term transport planning came to the forefront during the 1970s, not least because of the financial restraints imposed by governments and the 'energy crisis'. Historically the formalisation of short-term transport planning can be traced to the programme funded by the US Congress in 1968 called TOPICS (Traffic Operations Program to Increase Capacity and Safety) which was designed to: 'make traffic operation improvements on a systematic basis in accordance with an area-wide plan over a network of arterial and other major streets . . . and maximize the efficiency of the existing street system' (Gakenheimer and Meyer, 1978, p. 1). In September 1975, the joint regulations from the Federal Highway Administration and the Urban Mass Transportation Administration made transport systems management (TSM) 'a prerequisite for federal certification of local area planning processes' (ibid., p. 2) by requiring cities with populations over 200,000 to include short-term plans within a framework of longer-term improvements for transport (Keyani and Putnam, 1976, p. iii).

The wide-ranging nature of short-term planning makes it difficult to identify a common framework, but the systems approach is undoubtedly applicable (Lockwood and Wagner, 1977). The methodology of short-term planning is distinguished from other types of planning in three main ways. First, transport supply must be surveyed and represented for analysis in considerable detail. Second, as total travel demand changes relatively little in the very short term, elaborate forecasting models are less useful; behavioural models that are sensitive to policy changes are more suitable. Third, there should be more emphasis on experimentation and demonstration projects, which should be honestly monitored, evaluated and documented to provide information on the relevant features contributing to success or failure.

The objective of short-term transport planning is:

> to control in the short-term movement of people and goods on the urban transport network in a safe and efficient manner and in accordance with social concerns through the co-ordination in planning and implementation of the different elements of traffic management (OECD, 1977, p. 14).

There are many measures which can be employed individually, or, more usually, in combination, to achieve these aims: (a) traffic engineering techniques; (b) lorry routes; (c) traffic restraint; (d) parking controls; (e) bus priority; (f) public transport pricing and marketing; and (g) pedestrian schemes. A very useful reference source to consult is the survey of measures that have been implemented, evaluated and docu-mented in six European countries (May and Westland, 1979).

Traffic engineering techniques (Matson *et al.*, 1955) range from paint on the road to sophisticated computer-controlled traffic signals. Traffic control devices such as signs and lane-markings are the means to regulate, warn, guide or channel traffic. The need for this control at individual sites can be measured by the amount of delay to vehicles and pedestrians, or by the number of accidents that occur, and the effectiveness of such measures is judged by reductions in delay, safer driving conditions and less confusion to the motorist. Overhead signals, which designate the reversible lanes that can be used by time of day, are especially effective in adjusting transport capacity to match the directional imbalance of peak-period traffic flows.

Traffic signals (traffic lights) or mini-roundabouts are standard traffic engineering measures to produce an increased and more orderly flow of vehicles through busy intersections. Traffic signals are also warranted at uncontrolled junctions with bad accident histories, on major arterials where traffic from side-roads might otherwise make hazardous manoeuvres into or across the priority traffic stream, and at places where there is substantial pedestrian traffic, or where there is a need to protect a safe passage for pedestrians. In the CBD, town centres and major shopping centres, the linking or co-ordination of traffic signals will usually provide extra benefits over and above those obtained from the installation of controls at individual sites.

Specially designated lorry routes provide a way of controlling the movement of lorries in urban areas and of protecting residential amenity. Similar results are sometimes achieved by the proper sign-posting of through routes, or by making the alternative routes through residential areas less attractive to heavy vehicles; the advantage of such measures is that they become self-enforcing. This latter point is important because comprehensive lorry route networks are difficult to implement (Hasell *et al.*, 1978), despite the widespread public disquiet about heavy goods vehicles using unsuitable roads. Control over the delivery times of goods is very difficult to enforce in practice.

Traffic restraint is the recognition of the necessity to manage travel demand in central areas of cities. Traffic restraint measures, such as road congestion or deliberate pricing, attempt to reduce the amount of private travel by car and to make a more efficient use of the considerable investment in public transport (Bayliss and May, 1975). All methods of restraint should be supported by the control of city centre parking. Pricing mechanisms, such as supplementary licensing, a cordon of toll booths, permit schemes and metered vehicles, are aimed at *persuading* motorists to select alternative travel arrangements. Flexible employment hours of city centre workers, although not necessarily reducing the absolute amount of traffic, spread the peak demand for roads and public transport.

Whilst there has been favourable public reaction to the staggering of working hours (Selinger, 1977; TAU, 1977), governments, mindful of the political consequences, have been reluctant to implement traffic restraint schemes. A notable exception is in Singapore where an Area Licence Scheme was introduced in 1975. The key concept underlying the scheme is that a special daily or monthly supplementary licence must be obtained and displayed should a motorist wish to enter the designated 500 hectare restricted area between 7.30 and 10.15 in the morning. The Singapore government set a specific objective of reducing peak-hour traffic and careful monitoring showed that this target was more than achieved (Holland and Watson, 1978).

Parking control may be used both to reinforce restraint schemes and to form a major component of transport plans for busy activity centres (see Chapter 9). Measures include the complete prohibition or time restrictions on the use of on-street parking space, the pricing of parking, building codes that specify the amount of off-street parking in new developments, and the provision of peripheral parking facilities (park-and-ride) to encourage people to make the last part of the journey by public transport (Ellis, 1977). The restriction of parking, loading and unloading on main roads is essential when implementing urban clearways. Stationary vehicles reduce road capacity—three vehicles parked along a one-kilometre stretch of road reduces saturation flow by 16 per cent (Wells, 1975, p. 46).

Buses are the mainstay of public transport and although various measures giving buses priority in the traffic have long been in existence, it is mainly during the last decade that schemes have become co-ordinated to form an overall plan for urban public transport. A state-of-the-art report (NATO, 1976) describes the various schemes and assesses, with detailed case studies, their effectiveness. Priority measures include lanes reserved within the limits of the existing carriageway, bus-only streets, traffic signal settings which favour bus progression and traffic regulations which favour buses.

An alternative to tightening the constraints on private motoring or making it more costly is to set public transport fares lower. The case for reducing fares or having no fares at all is a strong one, especially if it is accepted that road pricing will be politically unacceptable.

Generally, fare pricing policies usually achieve less than the desired increase in patronage (May and Westland, 1979, p. 15), but promotional campaigns to market public transport services or to improve the image of public transport, coupled with better services and revised fare structures, have been partially successful in halting the decline in patronage.

Finally, pedestrian circulation is an important, if sometimes neglected, aspect of short-term transport planning, especially in the CBD and in major shopping centres. The ultimate aim is exclusive pedestrian streets or malls, but, in practice, delivery vehicles and emergency vehicles are given access, and in some schemes a further compromise is made to introduce buses and taxis. On many urban roads, the conflict between pedestrian circulation and vehicular traffic flow can be reduced, but never eliminated, with pedestrian crossings, overpasses, underpasses and special signal phasing at traffic lights that favour pedestrian flows (Stanford Research Institute, 1978). Pedestrian planning is considered in more detail in the next chapter.

Examples of short-term transport planning in Canberra are given in the next three sections. They include the planning and implementation of an area traffic control scheme, the planning and implementation of bus priority measures, and an urban transport study which investigates ways of managing travel demands in Canberra within a five-year time horizon.

8.2 Planning an Area Traffic Control Scheme

Area traffic control is essentially the method by which vehicular and pedestrian conflicts at road intersections are reduced to as low a level as possible. Traffic signals properly co-ordinated and controlled by computers allow the maximum amount of traffic to pass without enforced stops by assisting an orderly progression of vehicles while allowing for the competing claims of cross-street pedestrian and vehicular traffic.

In the planning of area traffic control schemes, the systems methodology is a very convenient framework to allow the main steps to be identified. Figure 8.1 illustrates the application of the systems approach to this type of planning. Elaborating on the items in the diagram, the following points should be noted.

(a) The multiple objectives of area traffic control and bus priority
    are to minimise delays and the number of stops for both vehicles

and pedestrians, to influence route choice so that drivers can avoid localised congestion, and to detect and give priority to emergency vehicles (police, fire engines, ambulances).

(b) The selection of an appropriate study area is important because traffic control devices are costly. Generally, this will be the CBD, other town centres or locations where traffic is very heavy and where intersections are close together.

(c) Comprehensive and detailed surveys are required of the existing roads and the prevailing traffic conditions. The data include street widths, distances between intersections, saturation flows, traffic flows, turning movements, travel times, delays, stopped time and queue lengths.

(d) The analysis is aimed at modelling traffic conditions for the proposed area traffic control scheme and how these conditions might be improved with alternative schemes.

(e) The performance of each alternative plan is evaluated by comparing the costs of installation and maintenance and the benefits arising from improved traffic flow.

(f) Systems design involves making decisions about the basic control strategy—either to use a set of pre-prepared traffic signal control plans that are implemented by time of day or to modify signals according to the prevailing traffic conditions from calculations made by an on-line computer. The component subsystem hardware necessary to do this includes: traffic detectors in the carriageway; communications both to transmit data recorded by the detectors to a central control and to transmit instructions back to the local traffic signal controllers; and a digital computer that processes the data and calculates appropriate signal plans and timings.

(g) Implementation obviously requires the political and financial support of governments together with the services provided by commercial firms manufacturing highly specialised traffic control equipment.

(h) Careful monitoring is necessary both to check that the system is functioning properly, to suggest modifications, if necessary, and to measure traffic conditions after the introduction of the scheme (Holroyd and Hillier, 1969, 1971).

Planning for a traffic control scheme in a study area involves determining the number and location of signalised intersections to be co-ordinated and the best way of linking and setting the signals.

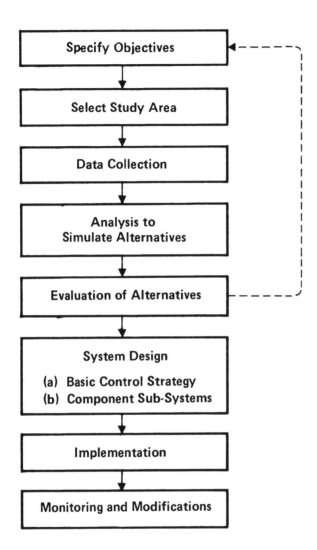

Figure 8.1: Steps in Planning an Area Traffic Control System

A helpful, and widely used, planning tool is TRANSYT, a digital computer programme that finds the best timings with which to co-ordinate traffic signals (Robertson, 1969a, b). It contains a mathematical model of traffic flow which enables the delay to vehicles and the number of stops to be estimated for different signal cycle times, green-time and offsets, and an optimisation algorithm which selects the green-times and offsets that minimise total vehicle delay and stops.

The study area transport system is represented in detail as a network of nodes and directional links (see section 2.3). Observed (or forecast) traffic enters this network from a set of 'external' links and the traffic flow pattern is given by a histogram (with the ordinate scaled to indicate vehicles per hour). The initial pattern of the traffic histogram entering the network is altered as it progresses through the network because some vehicles must stop at signalised intersections and because of different driver behaviour in terms of speed and vehicle spacing. This is represented by a model of platoon dispersion, the details of which are explained by Seddon (1972). In the TRANSYT programme, parameters control the amount of dispersion.

Given the pattern of traffic flow, the programme calculates deterministically the number of vehicles stopped at the signalised intersections and the total delay for an assumed cycle time, green-times and offsets. The stochastic nature of traffic is represented by an additional random term of 'total' delay. One summary of the traffic conditions on the whole network is the summation of the delays and stops on all links. This is called the 'performance index' (*PI*):

$$PI = \sum_{\ell=1}^{n} (D_\ell + \beta N_\ell) \qquad (8.1)$$

where

D_ℓ = the total delay in vehicle-hours, per hour on the ℓth link of the network;

N_ℓ = the average number of vehicle stops per second on the ℓth link;

n = total number of links in the network; and

β = stop penalty in seconds.

The size of the stop penalty may be chosen to correspond to what extent drivers are frustrated by having to stop at red lights. More complex performance indices may be specified to account for the stops and delays to different users of the transport network (vehicle classes, buses, pedestrians).

The TRANSYT optimisation model calculates the performance index for a given cycle time, green-times and offsets, and searches to see whether the index could be improved upon with different signal timings. Revised green-times and offsets will produce a different traffic pattern, involving a different component of delay and number of stops. The search for green-times and offsets continues until a *minimum* value of the performance index is found. This is the optimal signal timings for the signals under the prevailing traffic flow.

TRANSYT has been continuously refined over the years, and the inclusion of bus links allows signals to be set for bus priority (Pierce and Wood, 1977). The latest version, TRANSYT/7, is computationally more efficient, which is advantageous when analysing large networks (Hunt and Kennedy, 1978).

The minutes of a joint National Capital Development Commission/ Department of the Capital Territory meeting on 14 July 1977 reflect concern with increasing traffic congestion, especially with the longer-term build-up of employment in the central area. It was recommended that the policy of piecemeal 'isolated improvements' should be discontinued and replaced by an area traffic control policy, with the objectives:

(a) to optimise the performance characteristics i.e. minimise delays, stops, queue lengths and maximise capacities of general vehicles, buses and pedestrian movements with safety in the area as a whole and with some priority to public transport;
(b) to give priority to traffic on preferred routes and restrain undesirable traffic movements in accordance with the defined road hierarchy (NCDC, 1978b, p. 2).

A planning study, *Evaluation of Area Traffic Control Measures for the City Centre* (NCDC, 1977) investigated whether or not an area traffic control scheme could be justified in a study area in Canberra approximately 800 metres by 500 metres and containing, in 1977, six signalised intersections. The alternatives examined were: (a) signal co-ordination, improved cycle times and new phasing options; (b) bus priority measures (passive) achieved by signal plans favouring bus progression at intersections; (c) intersection improvements; (d) simple route control directing traffic away from bottlenecks by prohibiting various right-turning movements; and (e) alternative bus routes serving a proposed city bus interchange. Each alternative had four possible staging options, with the number of linked signalised intersections increasing from 3 to 5, then to 10, and finally to 16.

The analysis was conducted with the aid of the TRANSYT programme (NCDC, 1978c) which indicated the implications of each alternative when compared with the existing situation of isolated signals with their present phasing. The calculations were based on observed traffic during a morning and evening peak hour in 1976. The delay and stops were expressed in monetary units: travel time of car and truck drivers, bus passengers and pedestrians was valued at $A1.09 per hour (25 per cent of the average male wage rate); and each stop was valued at 1 cent for cars and trucks, and 3 cents per bus passenger.

Co-ordinated signals at 3 critical intersections produce large savings, which fall slightly for 5 intersections for site-specific reasons but then increase again as the scheme is extended to 16 intersections. The addition of simple route control and intersection improvements adds extra savings. Signal plans to favour bus progression is at the expense of other road-users, so, on balance, the savings are less. Restructuring bus routes and bus priority, together with signal co-ordination and route control, produces a saving of $A385,900 per annum, which is the largest saving of all.

An alternative way to judge the performance of the alternative schemes is to examine the likely traffic conditions. Table 8.1 shows the amount of delay, the number of stops, average vehicular speeds and air pollution at 16 intersections. The best scheme for motorists is signal co-ordination and route control because, compared with the existing situation of isolated signals, delays are reduced by 35 per cent, stops are reduced by 23 per cent, carbon monoxide emissions are reduced by 18 per cent and average vehicle speeds are increased by 34 per cent. Bus priority increases the number of stops for other vehicles and hence air pollution rises, but restructuring the bus routes together with bus priority improves efficiency.

Table 8.1: Estimated Performance Measures of Area Traffic Control Schemes in Canberra

Scheme	Delay hours/hr	Stops vehicles/hr	Average Speed kph Cars	Buses	Carbon Monoxide grams/hr
Isolated signals	453	23,540	20	18	160,929
Signal co-ordination	360	19,706	23	20	143,270
Co-ordination and diversion	293	18,012	27	22	131,570
Co-ordination, diversion and bus priority	295	22,032	26	23	140,130
Co-ordination, diversion, bus priority and optimal routes	263	20,454	26	26	137,030

Source: Nairn *et al.*, 1979, Figure 4, p. 7; and Akcelik, pers. comm.

In May 1979, it was announced that 12 sets of traffic signals in the city centre would be linked by telephones to a PDP 11 mini-computer at a cost of $A130,000 (*Canberra Times*, 3 May 1979). The system, expected to operate in about twelve months, is to be based on the Sydney Co-ordinated Adaptive Traffic System (SCAT), which was implemented in 1974 to control 150 signals in the Sydney CBD and had undergone further development (Sims, 1979). The normal mode of co-ordination is real-time, dynamic adjustment of cycle and off-sets in response to detected variations in traffic demand. However, the key feature of this system is its control flexibility—selecting either fixed signal plans or dynamic control, and breaking up large areas into smaller subsections, which can be linked or not according to prevailing traffic flows.

8.3 Bus Priority

The measures considered as suitable for giving buses priority in Canberra are bus lanes, busways and adapting traffic signals (section 8.2) to reduce the delay to buses. With-flow bus lanes, which are one of the most common forms of bus priority, are lanes reserved for bus use where the buses continue to operate in the same direction as the normal traffic. The advantage is that buses have free-running conditions and can bypass other vehicles at traffic signals or other bottlenecks. At signalised intersections, bus lanes are usually set back some 50 to 80 m from the stop-line to ensure an adequate flow of all vehicles through the intersection can be maintained. Generally, the effects of bus priority are to increase bus journey speeds by up to 5 kph and to decrease bus travel times by up to 25 per cent (NATO, 1976, p. 30), but case-study evaluations should be consulted to give a better idea of the advantages and disadvantages (Papoulias and Dix, 1978).

Contra-flow bus lanes are traffic lanes reserved for buses travelling in the opposite direction to the normal traffic flow. They are usually installed in one-way streets using a 3-metre-wide kerb lane, and may be divided from other traffic lanes by painted lines or physical separators. The beginning and end of such lanes often present the greatest geometrical design problems because other traffic must be redirected around each end but buses must be given ingress and egress (NATO, 1976, pp. 33-9). Apart from allowing speedier and more reliable bus services, a reason for introducing contra-flow lanes in one-way streets is to avoid physically splitting inward and outward bus routes and inconveniencing passengers.

Busways, which range from purpose-built, fully segregated roadways for the exclusive use of buses to a succession of inter-connecting bus lanes, allow continuous priority over substantial lengths of bus route. Segregated busways are more correctly considered as part of longer-term planning for public transport (Chapter 7), but in Canberra, where there is large-scale new development taking place, there is an opportunity to incorporate busways into a short-term action plan.

A feasibility study (Pak-Poy *et al.*, 1973) investigated the costs and benefits of alternative arrangements such as those discussed above to give buses priority on existing arterial roads connecting Woden Town Centre, City Centre and Belconnen Town Centre, a total distance of 17 km. The 'do-nothing' situation of buses in the mixed-traffic stream was modelled using traffic flow-dependent travel times calculated from Davidson's formula (equation (2.16)). For at-grade arterial roads a 'zero flow' travel speed of 48 kph was assumed, and a level of service parameter, λ, of 0.15 was selected; for grade-separated arterial roads, a speed of 56 kph and a level of service parameter of 0.10 were assumed. For the case with buses removed from the traffic onto a bus lane the link capacity in the formula was reduced to reflect the conversion of a carriageway lane into a bus lane. As the busway could be built entirely in the median, no loss of capacity was assumed.

Bus lanes reduce the total amount of travel time incurred by passengers by 42 per cent, but increase travel times for other road-users by 15 per cent for with-flow lanes and by 4 per cent for contra-flow lanes. For all users, with-flow bus lanes would increase the total amount of travel by 6 per cent, whereas contra-flow lanes would decrease the total amount of travel by 3 per cent, and the busway would decrease travel by 7 per cent. After weighing up the capital costs, the benefits and likely environmental impacts, the consultants recommended the busway on the median strip, where possible.

A section of this bus priority scheme was opened along Yarra Glen and Adelaide Avenue in 1975, but the with-flow bus lane located next to the median, and not the busway, was favoured. This decision was taken partly to protect a reservation for a future inter-town public transport system, which was still under consideration (see section 7.3). The median bus lane is effective because it carries mainly express buses and there are no signalised intersections along its length. By 1979, the priority measures had been extended to include pre-intersection bus lanes on Barry Drive, which connects Belconnen with the City Centre, and a busway through Belconnen Town Centre.

8.4 Transport Policy and Short-term Planning

The publication of a transport policy for Canberra in 1974 marked a shift in emphasis from long-term planning to devising ways of persuading commuters to use public transport to the central area and to other town centres, while leaving freedom of choice for other journeys. The *Canberra Short Term Transport Planning Study* (1977) was commissioned to investigate the consequences of policies favouring public transport and restraining private transport.

Because of the importance expressed in the policy statement of encouraging more people to use public transport, a disaggregate, behavioural modal-choice model was essential to the development of a 'new, more public-transport-oriented transport plan/programme for Canberra' (Daverin, 1977, p. 61). Three trip purposes were analysed: home-based work, home-based shopping and home-based social recreation. Five transport modes were defined: car driver ($m = 1$); car driver with one or more passengers ($m = 2$); car passenger ($m = 3$); bus passenger ($m = 4$), and walking ($m = 5$).

The travel demand model was structured as the probability of a traveller choosing one of these five alternative transport modes, given by the multinomial logit probability model (equation (3.29)). For example, the probability of driving alone ($m = 1$) to any one of the three trip purposes is:

$$P^r_{i(1)} = \frac{\exp\{f(X^r_{1(1)} + \ldots X^r_{n(1)})\}}{\sum\limits_{m=1}^{5} \exp\{f(X^r_{1(m)} + \ldots X^r_{n(m)})\}} \qquad (8.2)$$

where

$P^r_{i(1)}$ = probability of individual i choosing transport mode ($m = 1$) for trip purpose r;

$f(X^r_{1(1)} + \ldots X^r_{n(1)})$ = preference function for transport mode ($m = 1$) and trip purpose r; and

$f(X^r_{1(m)} + \ldots X^r_{n(m)})$ = preference function for the general case of transport mode (m) and trip purpose r.

The probability of choosing each of the four other transport modes is calculated in turn by substituing the appropriate mode-specific preference function in the numerator of the above equation.

These models were calibrated using data collected at 2,253 dwellings and relating to 25,319 individual trips (Bell and Symons, 1977). An appreciation of the essential structure of the model is obtained by considering, for example, the journey-to-work modal-choice equations.

The explanatory variables and parameters of the mode-specific preference function, $f(m)$, are listed below in notation form and these should be read in conjunction with Table 8.2, which describes the variables and gives their parameter values.

$$f(1) = \alpha + \beta_2 X_2 + \beta_3 X_3 + \beta_6 X_6 + \beta_7 X_7 + \beta_8 X_8 + \beta_9 X_9$$
$$f(2) = \alpha + \beta_2 X_2 + \beta_6 X_6 + \beta_7 X_7 + \beta_8 X_8 + \beta_9 X_9 + \beta_{10} X_{10} + \beta_{11} X_{11}$$
$$\qquad + \beta_{12} X_{12}$$
$$f(3) = \alpha + \beta_2 X_2 + \beta_3 X_3 + \beta_6 X_6 + \beta_9 X_9 + \beta_{10} X_{10} + \beta_{11} X_{11}$$
$$f(4) = \alpha + \beta_2 X_2 + \beta_4 X_4 + \beta_5 X_5 + \beta_9 X_9$$
$$f(5) = \alpha + \beta_1 X_1$$

Table 8.2: Calibration Parameters for Journey-to-work Modal-choice Behavioural Model, Canberra, 1975

Variable	Description	Parameter	Value
X_1	Travel distance (km)	β_1	−0.0900
X_2	Travel time by car (minutes)	β_2	−0.0182
X_3	Travel cost by car (dollars)	β_3	−0.0151
X_4	Walking time (minutes)	β_4	−0.0478
X_5	Waiting time (minutes)	β_5	−0.0453
X_6	Car availability: 1 car		4.972
	2 cars	β_6	3.420
	3+ cars		−0.165
X_7	Head of household: male		0.949
	female	β_7	1.282
X_8	Licensed driver	β_8	9.581
X_9	Sex of traveller: male		0.386
	female	β_9	0.926
X_{10}	Number of workers: 2		0.00089
	3(+)	β_{10}	0.00054
X_{11}	Number of full-time workers: 2		0.322
	3	β_{11}	−0.149
X_{12}	Flexitime: yes		−0.0627
	no	β_{12}	−0.0332
–	Constant term	α	−0.1624

Source: Based on CSTTPS, 1977c, Tables 3.4 and 3.5, pp. 20–1.

These equations are substituted into equation (8.2).

The planning application of behavioural travel demand models in Canberra was to examine changes to modal split stimulated by different transport policies and to trace their effects on different groups within the community. Variables in the model were changed to reflect each proposed transport policy, and the response of each individual was calculated and added up to form an aggregate response to the policy.

Modal-choice for all journey purposes was found to be insensitive to bus fares: bus patronage would either increase or decrease by only 1.5 per cent if fares were eliminated or if they were doubled. However, fare policy would have slightly more impact with the journey-to-work modal split. A doubling of fares decreases the observed modal split from about 10 per cent to 7 per cent, and a 100 per cent reduction in fares increases the modal split to about 13 per cent. Reductions in bus travel times to reflect the introduction of bus priority measures would have a negligible impact on altering modal choice. Reductions in access travel time by a half to reflect very convenient bus stop locations would increase modal split for the journey to work by about 2 per cent. Bus stops within a five-minute walk of all dwellings might double the number of social, recreational and shopping trips by bus from about 1,000 to 2,000 per weekday and very similar changes would occur if waiting times at bus stops were halved.

The effects of three parking charges on modal choice were investigated: a 'modest' fee of 20 cents per hour for long-stay parkers and 10 cents per hour for short-stay parkers; a 'severe' fee of 40 cents per hour, and 20 cents per hour, respectively; and a 'draconian' rate of $A2 per hour and $A1 per hour respectively. The likely effect of these parking charges on bus modal split for all journey purposes would be an increase to a maximum of about 14 per cent. The introduction of parking charges are likely to have a greater effect on altering modal split for the journey to work in Canberra. As illustrated by Figure 8.2, the relationship between modal choice and parking charges is non-linear, and most of the modal switching would occur for parking fees up to 75 cents per hour. Bus patronage would increase almost threefold to 25 per cent, car drivers travelling alone would drop from 55 per cent to 22 per cent, car passengers would increase to nearly 40 per cent.

The results of these analyses led the consultants to comment: 'Canberra must now make the transition from a laissez-faire transport policy environment to a "managed demand" environment' if it is to achieve the longer-term objectives of the Y-Plan (CSTTPS, 1977a, p. 91). The principal recommendations were the introduction of modest parking fees, the introduction of a surcharge on bus fares during the peak, the implementation of bus priority measures, the purchasing of 60 articulated and 280 ordinary buses and the development of a 'car pooling matching' programme at selected office complexes. Highways are given a lower priority and new investment in major road projects should be restricted to $A22.3 million up to 1982.

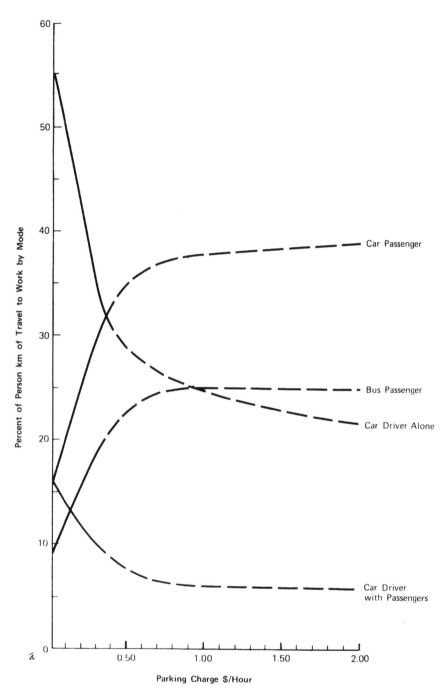

Figure 8.2: Parking Charges and Journey-to-work Modal Split, Canberra
Source: Based on CSTTPS, 1977a, Figure 4.8.

The implementation of measures to control parking in the city centre and improve public transport has not been without difficulty. From 1973, parking meters, with a 30-minute time limit and 5 cents for 15 minutes, and then voucher parking (10 cents for thirty minutes up to a two-hour limit) in car parks were introduced but these restrictions proved little deterrent to the use of cars, primarily because of the lack of policing of any illegal parking. A further 1,100 spaces in the Civic South, Petrie and Griffin car parks were converted to boom-gate controlled pay-parking in November 1977, with the principal objectives of discouraging long-stay parkers and providing accessible spaces at times of peak demand for short-stay parkers. Parking charges in 1979 were 20 cents for the first and second hours, 40 cents for the third, 60 cents for the fourth, and $A1 for the fifth and subsequent hours. Two months after opening, the car parks were underutilised because there had been a *redistribution* of both short- and long-stay parkers away from the parking areas controlled by boom-gates. There had been little noticeable effect on surpressing total demand (NCDC, 1978d, p. 16).

A series of problems confronted the Department of the Capital Territory in their efforts to improve bus services: there were delays in receiving delivery of new buses; there were cuts in public service staff ceilings that made the hiring of the necessary bus drivers difficult; and there was a shortage of qualified applicants. Most serious of all has been the reduction in the money allocated by the federal government to subsidise bus operations.

8.5 Summary

Short-term transport planning is concerned with more immediate action plans for making a better and more efficient use of existing urban transport facilities. The measures include standard traffic engineering techniques, parking controls, traffic restraint, bus priority, lorry routes and pedestrian schemes. The primary aim of short-term planning is to co-ordinate these measures into a comprehensive package of transport improvements.

Three examples of this type of planning approach were considered in more detail, and their applications to short-term planning in Canberra were discussed. The first described the planning of area traffic control, the second described bus priority schemes, and the third demonstrated the application of behavioural travel demand models to the formulation of urban transport policy, including public transport pricing strategies and parking charges. Some short-term transport planning measures, such as parking and pedestrian schemes, may also be classified as local area transport planning if their spatial impact is restricted, so this chapter and the next are in many ways complementary.

9 LOCAL AREA TRANSPORT PLANNING

Many traffic and transport issues are location-specific and therefore have a localised impact, but collectively they all add up and amount to a metropolitan-wide problem. As noted by Bayliss (1969, p. 21), the usefulness of the information generated by transport studies (of the type discussed in the previous four chapters) to spatially and topically more localised issues is limited. However, by treating topics at the appropriate scale, local area transport planning aims to supplement those planning activities at the wider, urban scale.

Local area planning is characterised by either proposals for entirely new developments or solutions for the existing urban fabric which are restricted in their areal extent—from individual sites up to residential neighbourhoods. Planning is usually for short-term horizons, and the plans therefore have much in common with those discussed in the previous chapter. Because the plans are localised and possibly influence the way development happens in the near future, the vested interests of planners, developers, governments, statutory bodies and the public become clarified and any conflicts are likely to be resolved only after protracted discussion and negotiation.

There is an enormous amount of material relevant to local area transport planning, so this chapter is very selective. At the local scale, walking is the universal transport mode and pedestrian planning is given prominence in the first section. Parking is also another local area problem and planning for parking in city centres and residential areas is considered in the second section. The traffic effects of proposed developments are explained in the third section. Finally, the planning and design concepts for streets and cycle paths in residential areas are introduced.

9.1 Planning for Pedestrians

Pedestrian traffic is far more fluid and adaptive than vehicular traffic, and for this reason pedestrian facilities have been designed intuitively (Morris and Zisman, 1962, p. 153) or have resulted more or less as byproducts from the formal architectural layouts of buildings. However, as noted by Hillman and Whalley (1975, p. 110): 'Walking and local

199

activity can be characterised as the bread and butter of daily travel, so the planning system should not overlook its responsibility to provide this staple diet.'

The initial task is to define the general aim of a pedestrian planning study (Wood, 1977, pp. 106-12). In the CBD and major shopping centres, the ultimate objective might be to cope with crowds by increasing the space for pedestrian circulation, and creating a continuous network of pedestrian-only streets. The elimination of vehicular traffic, the paving and landscaping of the area all contribute to providing a more pleasant, safer and more attractive environment. Compromises are necessary in practice and some pedestrian streets permit access to buses, taxis and delivery vehicles.

Data collection procedures are a special case of those methods outlined in Chapters 2 and 3, suitably modified in detail for the local scale. An inventory of transport supply involves measuring the configuration and dimensions of footpaths and other pedestrian facilities, and locating any obstructions to pedestrian flow. Measures of land-use intensity should relate to individual buildings and sites, not to zone centroids. Although home interview surveys provide an obvious means of establishing pedestrian traffic patterns, their disadvantages are that the origin and destination of journeys are not always located precisely enough and walking trips, especially short ones, are under-reported (Rigby, 1977, p. 2). Other survey methods are field-counts of pedestrians crossing an imaginary cordon and time-lapse photography.

The main analytical task is to determine whether the space provisions in existing or proposed developments are satisfactory for pedestrians and sensible analyses can be conducted by considering space requirements for humans at rest and in motion, and pedestrian walking speeds in crowded and uncrowded situations. A good measure of the adequacy of facilities is whether people can walk at unrestrained speeds. In Western countries average walking speeds on the flat, or on grades up to 5 per cent, are 80 m per minute (4.8 kph). The equivalent average (horizontal time-mean) speed walking up stairs is about half that walking on the flat.

Another measure of adequacy is the space taken up by a person standing. Body depths and shoulder breadth are the dimensions commonly used for designing pedestrian spaces and facilities and a design template corresponding to an ellipse of 61 cm × 46 cm is recommended (Fruin, 1971, p. 20). Psychologists observe that when people crowd into spaces smaller than $0.25 \, m^2$ per person personal discomfort is apparent.

Space required for locomotion is largely dependent on the length of strides that people take and on reaction time to avoid colliding with other pedestrians. Above 2.3 m² per person, normal walking speeds are possible, but as space is reduced average walking speeds drop until a jam occurs with 0.23 m² per person.

In analysing the adequacy of pedestrian facilities two important relationships must be appreciated: (a) average walking speed is a function of pedestrian traffic density; and (b) maximum pedestrian traffic flow is related to traffic density:

$$Q = V/A \qquad\qquad (9.1)$$

where

Q = pedestrian traffic flow, measured in the number of pedestrians per metre width of walkway (or stairs) per minute;

V = mean walking speed, metres per minute; and

A = area per pedestrian in m² (as a measure of traffic density).

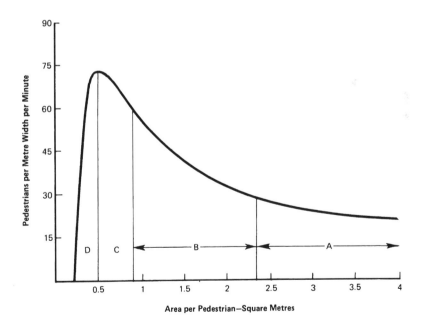

Figure 9.1: Pedestrian Flow and 'Level of Service' Concept
Source: Based on Fruin, 1971, Figure 4.1, p. 78.

This relationship is clarified by referring to Figure 9.1, which illustrates directional pedestrian traffic flow on the vertical axis of the graph and the area per pedestrian on the horizontal axis. The maximum flow that can be sustained over a substantial period of time is about 72 persons per metre width per minute on a footpath (about 50 persons per metre width of stair). Interestingly, minor reverse traffic flow on footpaths has little effect on reducing this flow.

The three vertical lines on Figure 9.1 define four 'levels of service' to the pedestrian, which can be used by planners to judge how satisfactory pedestrian facilities are, or might be (Fruin, 1971, Chapter 4). Table 9.1 provides a summary description of pedestrian traffic conditions experienced at each 'level of service'. A high standard of design (A or B) implies that pedestrians are free to walk normally, to overtake dawdling pedestrians, and to avoid contact or conflict with cross or reverse flow movements.

When making calculations for any real situation the 'level of service' should be computed from the *effective* space for circulation (O'Flaherty and Parkinson, 1972). Field surveys can determine the loss of width due to obstructions, but in the absence of authentic data, Fruin (1971, p. 44) suggests 0.5 m be subtracted for shop-window gazing and a further 0.7 m subtracted for kerbside obstructions, such as telegraph poles, lamp standards, parking meters or queues at bus stops.

Table 9.1: 'Level of Service' Concept for Footpaths

Level of Service	Traffic Flow (ped/m/min.)	Area (m²/ped)	Pedestrian Traffic Conditions
A	≤30	>2.3	Free flow
B	30 – 55	2.3 – 0.9	Individual walking speeds and passing manoeuvring restricted; pedestrians involved in reverse flow or cross movements severely inconvenienced
C	55 – 70	0.9 – 0.5	Restricted walking speeds and frequent adjustments of gait with some shuffling; insufficient room to by-pass slower movers; extreme difficulties for pedestrians attempting reverse or cross manoeuvres
D	≥70	<0.5	Erratic flow, with sporadic forward movements; unavoidable body contact; reverse and cross movements impossible.

Source: Based on Fruin, 1971, pp. 74–8.

The remaining steps of the systems approach to pedestrian planning are plan preparation, evaluation and implementation, and these are not discussed because real case studies are more informative (Elkington *et al.*, 1977; Brambilla and Longo, 1977).

9.2 Planning Parking Facilities

Despite attempts in major activity centres to reduce the intrusion of motor vehicles and to encourage more people to take public transport, the parking of motor vehicles remains a major local area transport problem. The general aims of planning parking facilities are fairly broad, and sometimes the objectives may be contradictory. For example, controlling the number of parking spaces in city centres as part of any traffic restraint programme is a different philosophy to the one of locating ample parking convenient to the motorists' ultimate destination. In formulating a comprehensive parking policy for the CBD, town centres and major shopping centres (O'Flaherty, 1974, p. 129; Armitage, 1977) some of the considerations are: (a) to balance the amount of parking with available road capacity; (b) to maximise road capacity by controlling kerbside parking; (c) to strike a fair balance of parking supply amongst different users, such as delivery vehicles and customer parking; (d) to reduce the visual intrusion of parking structures or parked vehicles; (e) to ensure that the traffic flow from a car park does not overload adjacent streets; and (f) to enhance the commercial interests of retailers and business.

Because some form of parking control is needed almost everywhere in central areas, policies often extend to the older inner-city residential areas which often have a mixture of non-conforming land uses, and where there is insufficient space for off-street parking. The aims might be to suppress the on-street parking demand by introducing parking meters or other pricing mechanisms, or to give residents priority by issuing parking permits (Collins and Pharoah, 1974, pp. 468-508; Elliot, 1977). Residential amenity can be protected by banning overnight parking of lorries on residential streets (GLC, 1976, p. 91).

All policies for parking should be based on accurate information about how satisfactory or otherwise existing facilities are functioning. Parking surveys (Burrage and Mogren, 1957, pp. 60-75) establish the location of available parking supply and the spatial pattern of parking usage. The four main kinds of information collected from site inspections, and located on a suitably scaled map, are: (a) the number of parking spaces; (b) parking regulations, such as loading and unloading zones, time restrictions and other types of control (meters, discs displayed on vehicles); (c) the type of parking facilities—on-street, surface, multi-storey, underground; and (d) any time limits or parking charges.

True parking *demand* is difficult to establish because of the various constraints imposed by the locational shortages in parking supply, but field surveys can determine parking usage. The important data to collect from site inspections are the arrival and departure times of vehicles and the accumulation of vehicles parked by time of day. Short-stay parkers are sometimes missed by an observer cruising around and revisiting each parking spot at periodic intervals, and suitable correction factors can be applied.

An important factor to analyse in the operation of any car park is the average duration of time vehicles are parked because the turnover of spaces is as important as the number of spaces supplied. The mean duration of stay and the number of arriving vehicles per unit time represent the traffic load placed on a parking facility:

$$A = Q\bar{T}_d \qquad (9.2)$$

where

A = traffic load;

Q = number of vehicles arriving per unit time; and

$\bar{T}_d$ = mean parking duration in time units.

For example, suppose the mean parking duration was 30 minutes and vehicles arrived at a rate of 100 per hour, then the traffic load on the car park would be 50. The traffic load is important as a theoretical concept because when the load is too great the car park becomes full, and drivers must cruise around or look for an alternative place to park.

Some useful calculations involving the traffic load and the number of parking spaces can be made if a car park is conceptualised as a multiple channel queueing system with M-parallel channels that correspond with M parking spaces (Blunden, 1971, pp. 57-9). The steady state 'probability of rejection' (being turned away) increases with the traffic load and decreases with the number of spaces:

$$P_r = \frac{A^M/M!}{1 + A + A^2/2! + \ldots A^M/M!} \qquad (9.3)$$

where

P_r = probability of rejection.

Table 9.2 evaluates this formula for selected values of *A* and *M* because it is a lengthy computation.

Table 9.2: Probability of Rejection for Selected Traffic Loads and Parking Spaces

Traffic Load	Number of Parking Spaces				
(*A*)	*M* = 1	*M* = 5	*M* = 10	*M* = 50	*M* = 100
1	0.50	0.00	0	0	0
2	0.67	0.04	0	0	0
3	0.75	0.11	0	0	0
4	0.80	0.20	0	0	0
5	0.83	0.28	0.02	0	0
10	0.91	0.56	0.21	0	0
50	0.98	0.90	0.80	0.10	0
100	0.99	0.95	0.90	0.51	0.08

Source: Based on Rallis, 1967, Table 1a, p. 21; and Blunden, 1971, Table 3.1, p. 59.

The 'probability of rejection' is one theoretical measure of the performance of a parking facility under differing traffic conditions. The main implication of the results shown in Table 9.2 is that larger parking lots operate more satisfactorily than ones with few spaces: when there is only a small cluster of spaces the 'probability of rejection' is considerable, but as the number of spaces increase, the 'probability of rejection' is quite small, even when the traffic load approaches the total number of parking channels.

The provision of lorry loading and unloading docks and the number of reservoir (waiting) spaces on the site of commercial, retail or industrial premises can be analysed in much the same way using multiple channel queueing theory (Watters, 1972). The calculations can be expressed as a series of design graphs which indicate the number of docks and reservoir spaces for various traffic loads and 'probabilities of rejection'. Figure 9.2 gives an example of a site with one reservoir space. The vertical axis is labelled the 'probability of rejection'—that is, a lorry finds all docks *and* the one reservoir space occupied—and the horizontal axis is labelled the traffic load, which in this case is the ratio of the average arrival rate to the average rate of loading or unloading. It is up to the discretion of the planner to select the design standard for the number of docks (with its development cost implications), bearing in mind the potential traffic problem that might be caused by a lorry parked on the street waiting for a dock to become vacant. The number of kerbside parking spaces required for commercial vehicles unloading in shopping areas can also be treated in a similar way.

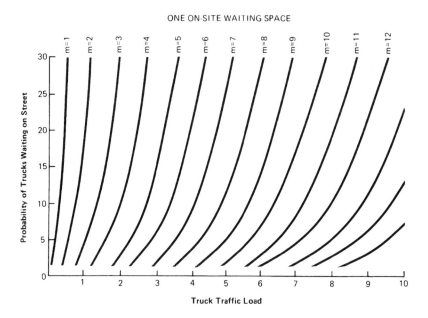

Figure 9.2: Design Chart for Planning Loading and Unloading Docks
Source: Based on Watters, 1972, Figure 2B.

Forecasting parking needs is fairly complicated because of the interaction which gives an equilibrium between the demand for and the supply of parking spaces. Generally, it is difficult to consider individual parking facilities other than on a short-term basis and forecasts are usually made for each central area traffic zone. Figure 9.3 gives the broad steps to follow in making forecasts of parking requirements at a strategic level.

The right-hand side of Figure 9.3 is concerned with future parking demand. This demand is derived from urban travel demand model forecasts: the zonal number of vehicular trip-ends is the main input to a parking model, which converts the number of zonal trip-ends into parking requirements based on parking duration and arrival rates. For an indication of on-street parking requirements in inner-city residential areas forecasts of household vehicle ownership are necessary. Forecasts of modal split and car occupancy for the journey to work are necessary in an assessment of parking requirements at non-residential land uses. The use of generalised cost, including terminal parking penalties in the specification of transport supply, allows the effect of parking charges on modal split to be assessed (see section 8.4).

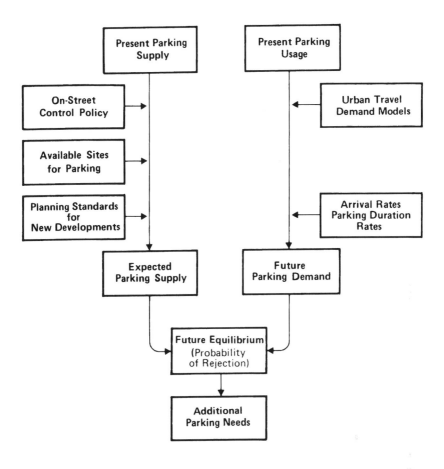

Figure 9.3: Main Steps in Forecasting Parking Requirements

The left-hand side of Figure 9.3 indicates that the supply of parking is dependent on: (a) the existing provisions *minus* any spaces lost through the control of on-street parking; (b) the availability of sites with potential for parking facilities—which is tied closely to land-use policy; and (c) parking standards or building codes enforced in the approval of development applications (Bayliss, 1969). A parking code is the main mechanism for controlling off-street parking supply as town centres are redeveloped. Parking codes are often 'rule of thumb' measures (Shoup and Pickrell, 1978) and vary widely from place to place.

In Figure 9.3 the future parking demand and the expected parking supply are drawn together to indicate the additional parking needs, if any. A guide on how to compile a series of maps on the existing and proposed parking arrangements is outlined by O'Flaherty (1974, pp. 166-77).

9.3 Traffic Effects of Land-use Developments

One genuine local area transport planning issue is the traffic implications of any proposed development, especially if the site has substantial off-street parking and a main road frontage. Traffic criteria are often used by planning authorities in assessing development applications. Before planning approval is granted, the two main questions to resolve are: (a) how much traffic will be generated by the new development? and (b) is there sufficient space capacity on the surrounding streets, or will vehicular manoeuvres into and out of the site interfere with the free flow of traffic on the main road? A spectacular example of this is in the planning of casino-hotels in the seaside resort of Atlantic City, New Jersey (Lee and DePhillips, 1979).

The likely amount of vehicular traffic generation can be calculated in two different ways. The first is to collect data for similar types of development and to derive an empirical relationship between traffic generation of a site and measures of land-use intensity on that site using regression analysis (Chapter 3, section 3.2). The value of land-use intensity anticipated for the proposed development is substituted into the equation and an estimate of traffic generation is obtained (Field *et al.*, 1979).

Alternatively, the average duration of time that vehicles are likely to spend at the proposed development can lead to simple calculations of the *maximum* traffic generation of the site (Blunden, 1971, pp. 229–31). By taking measurements of the time that it takes to be served at similar types of development to the one proposed (for example, at petrol service stations, drive-in liquor stores or fast-food outlets) or the time that vehicles are parked (for example, in car parks associated with each development) the mean 'service time' is established. The turn-around rate for each space is a function of the service time:

$$\mu = 60/\bar{T}_s \qquad (9.4)$$

where

μ = turn-around rate, vehicles per hour; and
$\bar{T}_s$ = mean 'service time', in minutes.

For a multiple number of spaces, such as the number of petrol pumps, the number of serving outlets or the number of parking spaces, the maximum possible vehicular traffic production per hour is:

$$Q_{max} = \mu M \qquad (9.5)$$

where

Q_{max} = maximum possible number of vehicles per hour produced by the site;

M = total number of 'spaces' or 'service channels'; and

μ = turn-around rate of vehicles per hour.

Total traffic generation of the site (trips in *and* out) is double the figure obtained from equation (9.5).

In assessing the traffic implications of isolated development there is considerable merit in examining the 'worst possible case' when the peak traffic flow on the frontage road coincides in time with the *maximum* flow into and out of the development. If such a situation is tenable no further evaluation is required. However, for major development applications such as for office complexes or for multi-storey car parks, the 'phasing' between the main road traffic and the traffic flows into and out of the development should be determined with accuracy from field surveys.

The important question is whether or not the entry-exit movements associated with a proposed development overload the surrounding road system. Appropriate calculations can be made with a theoretical formula which indicates the number of vehicles per hour which can be 'absorbed' satisfactorily by a priority traffic stream (Blunden, 1971, p. 71):

$$Q_a = \frac{Q \exp(-Q\bar{T}/3600)}{1 - \exp(-Q\bar{t}/3600)} \qquad (9.6)$$

where

Q_a = the number of vehicles per hour 'absorbed' by a priority traffic stream;

Q = traffic flow in vehicles per hour for the priority traffic stream;

$\bar{T}$ = mean acceptance gap in seconds for the first vehicle to enter a headway; and

$\bar{t}$ = mean 'follow-up' acceptance gap in seconds.

Exit movements of vehicles from the site must wait until a suitable gap occurs in the passing traffic. The headway required for a stationary vehicle to merge into a traffic stream is the acceptance gap, $\bar{T}$, in equation (9.6). When a number of vehicles are queued waiting for gaps, and a sufficiently large gap opens in the traffic stream, the second and subsequent vehicles accept a smaller 'follow-up' acceptance gap, $\bar{t}$.

The horizontal and vertical alignment of the road and any obstructions in the vicinity of the site determine sight-distances from the exit point, so the character of acceptance gaps should be established empirically. However, the following parameters are suitable for approximate calculations. In the case of left-turn exits the value of $\bar{T}$ is usually taken as 4 seconds, and the priority traffic flow is measured for the inside lane. For right-turn exits the value of $\bar{T}$ is 5 seconds, and the priority traffic flow is measured for the total flow in all nearside lanes plus the flow in the centre lane of the offside traffic (Blunden, 1966). 'Follow-up' acceptance gaps are 2 seconds for left-turn exits and 3 seconds for right-turn exits.

The application of these concepts to examining the traffic implications of a proposed land-use development are illustrated with a simple example. The site, by a two-lane road, is expected to generate 600 vehicles per hour in the busy period comprising 60 per cent left-turners and 40 per cent right-turners. The peak traffic flow on the lane adjacent to the proposed development is 800 vehicles per hour and 1,000 vehicles per hour in the opposite direction. The absorption rate for the adjacent lane is 916 vehicles per hour ($\bar{T} = 4; \bar{t} = 2$) and the number of left-turning vehicles is only 360 per hour so there is plenty of 'spare capacity'. However, the absorption rate for both traffic streams is 190 vehicles per hour ($Q = 1,800; \bar{T} = 5; \bar{t} = 3$), whereas the number of right-turning vehicles is 240 per hour. If the development were approved, then right-turn exiting vehicles would suffer delays.

9.4 Streets in Residential Areas

A planning topic of considerable importance, but one which does not fit conveniently into a systems framework, is residential streets and cycle paths, not least because about one-fifth of any residential subdivision is devoted to streets, and about two-fifths of the total development costs are attributed to road works and drainage (Wills, 1976, pp. 45-6). This section is more concerned with design and the broad aims of planning residential streets and cycle paths.

The dominant aim of planning is to design urban roads with a clear functional hierarchy (Tripp, 1942, p. 54; Buchanan, 1963, pp. 43-4): motorways and arterial roads are those major routes specially designated to canalise large volumes of traffic moving longer distances from one locality to another; other roads and cycle paths should provide for shorter journeys to schools, shops and recreation, and give access to residential properties. By creating a residential street pattern that discourages through traffic, residential amenity—traffic noise, accidents, delays to pedestrians crossing streets—can be protected.

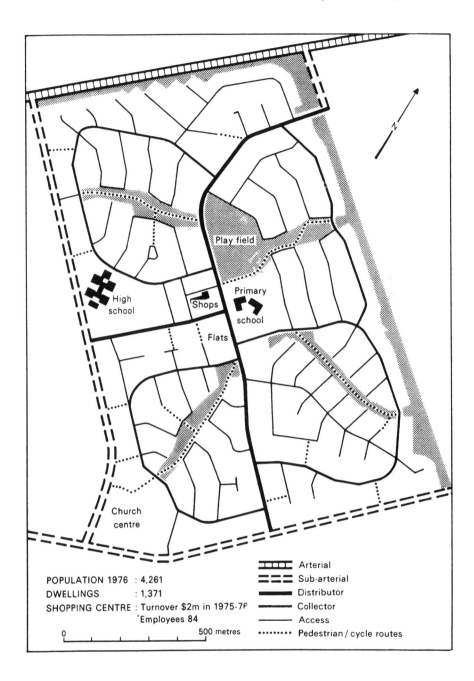

POPULATION 1976 : 4,261
DWELLINGS   : 1,371
SHOPPING CENTRE : Turnover $2m in 1975-76
                  'Employees 84

0   500 metres

▭▭▭▭ Arterial
▬▬▬ Sub-arterial
▬▬▬ Distributor
▬▬▬ Collector
───── Access
••••••• Pedestrian / cycle routes

Figure 9.4: Road Hierarchy in the Canberra Suburb of Watson

Figure 9.4 illustrates the road hierarchy concept using as an example the Canberra suburb of Watson. Major roads surround but do not pass through the neighbourhood. Almost all of the roads which are purely residential are in the form of loops and culs-de-sac, thus inhibiting through traffic and vehicle speeds. The design provides for maximum safety and convenience of pedestrians and cyclists, especially to the local shops and primary school, and there is a considerable degree of segregation of pedestrian, cycle and motor traffic ways (NCDC, 1975, pp. 43-50).

In the design of residential street layouts a set of principles to follow is helpful (DpT, 1977). The following list of guidelines, taken from *Road Safety Guidelines for Town Planning* (DT, 1978, pp. 2-4), is typical of much current thinking in the treatment of new residential developments:

(a) a hierarchy of streets should be provided;
(b) each street should only intersect with streets of the same class or in the class immediately above or below it in the hierarchy;
(c) direct access to arterial roads should be restricted to collector or distributor roads, and these intersections should be T-junctions and signalised if warranted;
(d) direct access to private property from arterial roads should not generally be permitted;
(e) each class of road should be designed to a recognised standard, related to function and traffic speed;
(f) local streets should be designed to limit vehicle speeds;
(g) collector streets should not provide continuous through routes for traffic between arterials;
(h) local/local, local/collector and collector/collector should be T-junctions, but acute angle Y-junctions should be avoided;
(i) intersections should provide adequate sight-distance;
(j) the street layout should be such that the traffic generated by local streets and carried by the collectors should be within the range of 2,000-2,500 vehicles per day; and
(k) pedestrian and cycle facilities should be included.

In practice, street planning relies heavily on the creative skills of the designer using the characteristics of the site as established by proper field investigations and this is a traditional part of town planning (Lynch, 1971, Chapters 2 and 4).

A controversial issue is the approriate right-of-way widths and their division into carriageway pavements, footpaths and grassed verges. In practice, geometrical design and construction standards vary widely (Holton and Pattinson, 1976).

As a very general guideline only, Table 9.3 is indicative of the NCDC's current planning approach with respect to the minimum dimensions of cross-sections of different categories of road in Canberra. Design speeds are 55 kph for distributors, 50 kph for collectors and 40 kph for access streets.

Table 9.3: Minimum Residential Road Cross-section Dimensions, Canberra

Type of Road	Carriageway Width (metres)
Cul-de-sac, when 75 m long	4.5 to 6.0
Cul-de-sac, when 150 m long	5.0 to 6.7
Short loop	5.0 to 6.7
Long loop	5.5 to 7.3
Collector	5.5 to 7.3
Collector, serving high-density housing	5.5 to 10.3
Undivided distributor	5.8 to 12.8
Divided distributor	dual 6.7 m pavements

Source: NCDC, 1975, p. 52.

Planning alterations to roads, especially street closures, in established residential areas is a contentious topic, not least because of lobbying and protesting of local residents likely to be affected by such schemes. The measures that can be taken to change street layouts are well established. Traffic engineering measures include reducing carriageway widths, mid-block closures to create culs-de-sac, reconstructed approaches at cross-roads, diagonal closures and traffic control devices such as stop signs (Vreugdenhil, 1976).

The main considerations in planning bicycle facilities (Hudson, 1978) are to provide a continuous network of safe and direct routes, which may even include transportation of bicycles on established public transport. However, satisfactory arrangements for leaving bicycles unattended at the destination are also essential if cycling is to be encouraged as a serious transport mode and consideration should be given to parking facilities in activity centres and at intermodal interchanges. In new residential developments, or in recreational areas, cycle paths can be located in attractive surroundings to encourage cycling as a leisure pursuit, such that the conflict with both vehicular traffic and pedestrians is minimised. Dimensions of cycle paths are less controversial than street widths because a single cycle-lane can accommodate from 850 to 1,000 cycles per hour in both directions and capacity is not an issue. Cycle paths are normally built 3 to 4 m wide, with gentle alignments and easy grades.

In Canberra, there is an extensive programme for the development of a metropolitan-wide cycle path network. The first cycle path, constructed in 1973 from the Australian National University to Dickson in North Canberra, has been extended northwards. Recently completed cycle paths link the inner-city suburb of O'Connor with the Belconnen Town Centre, and South Canberra with Woden Town Centre. Recent monitoring of the use of cycle facilities in Canberra has found that the number of cycle trips has increased substantially: on the Dickson cycle path trips during the morning peak two hours had doubled in three years.

9.5 Summary

A chapter entitled 'local area transport planning' has been included because there are numerous location-specific traffic and transport problems which are sometimes overlooked in conventional transport studies and which, if tackled correctly, could contribute to an improved urban environment. Selected examples described include planning pedestrian and parking facilities, the planning of streets and cycle paths in residential areas, and the localised traffic effects of isolated land-use developments. Inevitably, some topics have been excluded, and others glossed over. More research is required to develop more suitable methods of local area transport planning.

10 OVERVIEW AND FUTURE DIRECTIONS

The aim of this book has been to introduce the methodology of urban transport planning by explaining theoretical developments and their practical applications, and not to offer a critique of the process. But criticism is an essential part of scientific method because new techniques emerge and procedures are modified where weaknesses are exposed. Already, the reader may have formed impressions about strengths and deficiencies of transport planning methodology, or may have thought of suggestions for improving it, but, if not, this final chapter is a guide to developing a more critical appreciation of transport planning.

A theme which re-occurs throughout the text is the systems approach, and this should be defended. It is argued that the systems approach is a suitable methodology for the analysis of large, complex systems. Above all, the systems approach provides a very convenient framework by reducing the intricate details of the transport planning process to a few major steps such as problem definition, data collection, analysis and forecasting, plan evaluation and implementation.

Systems analysis as a legitimate study area has been severely criticised (Berlinski, 1976; Sayer, 1976), not least because of its ideological implications. According to Healey (1977, p. 204) the transport planning establishment is 'dominated by a hierarchist ideology rooted in a one-way causal paradigm' which relies on positivist procedures, such as inferring the future directly from the past with the assistance of mathematical systems models. Furthermore, such planning procedures are based on the concept of the expert: the community is viewed as ignorant and incapable, and consequently is kept uninformed by the planners until the important decisions have been taken.

The power of this criticism can be largely defused in two ways. The first is probably untenable in practice because it assumes that planners are able to work genuinely in the 'public' interest, however defined. Although it is generally assumed that technocrats are, or should be, objective and politically neutral, 'the idea of political neutrality demands that individuals have no values and make no choices. But no planning information ... is more political—and moral or immoral—than the traffic planner's choice of what to feed into his computer' (Sandercock, 1975, p. 128).

The second way is for more public involvement in the planning process—in the UK, public participation is a statutory requirement of the development plan process. In the past, any lack of public participation in transport planning is a failure not of the systems approach but of the way that it has been applied. In fact, in *Principles of Urban Transport Systems Planning*, it is stated clearly that planning is a continuous process involving a dialogue between governments and the public, and that the systems approach is an extension of a simple frame-work of government-community interaction (Hutchinson, 1974, p. 6).

The role and value of mathematical models in the planning process has been at the centre of much lively debate. One school of thought assumes that there is no substitute for common sense and any attempt at pseudo-scientific justification for a transport plan is an unnecessary waste of effort. For example, it has been argued that the broad results from models are so general as to be self-evident and that specific results are so inaccurate to be virtually useless for practical planning (Robbins, 1978, p. 35).

Another school of thought is that urban problems can be solved by computer-based analytical models, but this, too, is an extreme view. The computer makes it practical to handle large volumes of data many orders of magnitude larger than before and to analyse these data using more sophisticated techniques, but it is easy to become preoccupied with model-building for its own sake and distracted from the realities of planning problems. Practical experience with complex computer models has been one of frustrating delays (Hemmens, 1968) and expensive experimentation with little tangible results to show for the effort involved (Drake, 1973).

Many criticisms against mathematical models can be largely eliminated provided that their assumptions are acknowledged and their results are interpreted sensibly. The analyst is less likely to be hood-winked by the apparent illusion of accuracy of transport models in reproducing base-year traffic if the theoretical structures of the models explained in Chapters 2 and 3 are studied carefully and if the goodness-of-fit tests are performed. Modelling accuracy becomes less of an issue when results are used to examine and compare alternative plans on an internally consistent basis.

Sequential models are only as strong as the weakest link in the chain, and a justified criticism of traffic forecasting models is the uncertainty surrounding the future level of social and economic activity and its spatial distribution. Exogenous land-use variables are necessary inputs to transport models: traffic forecasts are 'based on an ambitious range of assumptions and forecasts . . . which any specialist in economic or social forecasting would regard with severe caution' (Solesbury and Townsend, 1970, p. 69).

The purpose of quantitative analysis is to present to decision-makers information about the implications of different courses of action. Generally, there has been an unsuccessful marriage of economic evaluation methods with the transport planning process (Troy and Neutze, 1969): only about one-half of the urban transport studies conducted in Australia presented any formal evaluation of their plans (Black, 1975, pp. 324-7). As noted by Duhs and Beggs (1977, pp. 228-9), 'the relationship of economics to the plans has all too often been rather like that of Cinderella to the ball—lucky to get in, and packed off home before the real event was scheduled to start.'

Even when a satisfactory economic evaluation has helped to make transport investment decisions, the effect of transport on the distribution of wealth within the urban area has largely been ignored (Neutze, 1978, p. 118). Mathematical models can be stratified by social group, but in practice this increases the analytical task considerably. Better methods of impact analysis that embrace the distribution of costs and benefits are required and more attention needs to be afforded to energy and environmental considerations when evaluating transport plans.

Transport planning studies, especially of the kind described in Chapter 5, have been widely criticised and most transport planners are aware of their shortcomings. The major weaknesses of conventional land-use—transport planning studies, ranked according to the frequency that they were mentioned by Australian planners in a survey by Leveris (1975) are:

(a) the acceptance of a given land-use arrangement and the failure to analyse alternatives;
(b) a low degree of implementation;
(c) lack of public participation;
(d) inadequate treatment of public transport;
(e) too costly and time-consuming modelling; and
(f) neglect of social and environmental impacts.

Failure to consider the long-term traffic implications of alternative land-use plans, as noted in Chapter 5, reflects an inadequate understanding of the systems planning approach. However, the view has been expressed, especially in those countries with limited statutory planning controls, that any analysis of a future land-use pattern other than an extrapolation of existing conditions is a waste of time (Levin and

Abend, 1971, pp. 78–89). Even when alternatives have been examined, the analyses have failed to show significant differences in travel patterns or substantial differences in the costs of providing transport facilities (Boyce *et al.*, 1969, pp. 84–8). But the credibility of any longer-term planning exercise is enhanced only when a range of plausible alternatives are examined, as in Chapter 6.

Many urban transport studies have recommended extensive motor-way networks which have never been built, so generally there has been a low degree of plan implementation. Public opposition to motorways has become a political issue (Thomson, 1969) and many governments have abandoned such plans, especially for roadways in inner suburbs. Government expenditure of all kinds, including investment in urban transport, has been restricted in the 1970s and is unlikely to be available on the scale necessary to construct massive road programmes or new fixed-track public transport systems. In times of financial restraint, making a better use of existing transport facilities or reducing the deficit of public transport with short-term, incremental planning (Chapter 8) becomes the only practical option.

The need for greater community involvement in the transport planning process is widely recognised, but it is not obvious how to incorporate public participation–'methodologically, participation raises the problem of simply how it is to be structured in the planning process' (Healey, 1977, p. 220). Local area transport planning (Chapter 9) brings into sharp focus conflicts of interest and provides the opportunity for genuine local community participation. The task of the planner is to identify the relevant issues and value positions of each group and to indicate the consequences of alternatives to the decision-makers and to the public rather than making technical decisions on behalf of society.

Public transport was given relatively little attention in some urban transport studies, which were, in many cases, commissioned by highway authorities during a period of rising car ownership, declining patronage on public transport, and optimism towards future energy sources. This reflects government transport policy which has largely treated urban transport modes independently. In the UK, prior to 1968 most of government help towards urban transport expenditure was concentrated on roads, but a more balanced approach followed the 1968 Transport Act when infrastructure grants were made available for railways, busways, bus priority schemes, new buses and interchange facilities (Starkie, 1976, pp. 36–7). The planning studies described in Chapters 7 and 8 suggest how any bias towards road planning may be removed from transport planning methodology.

Transport planners should be conscious of the costs and time involved with computer-based modelling exercises. The experience in the United States, according to a quote by Drake (1973, p. 1), is summarised thus: 'Never before in the history of human conflict has more money been spent, by more people with less to show for it.' It is probably more true to say that because of the slavish attention to detail which modelling requires, 'many of the most able engineering brains are directed from creative design work and others are trapped in the drudgery of preparing what is little more than "computer fodder" ' (Robbins, 1978, p. 34). Simplified analytical procedures, along the lines outlines in Chapter 6, represent one way of overcoming this problem by allowing more alternative land-use–transport plans to be examined efficiently (Plowden, 1967; Schofer and Stopher, 1979).

Finally, concern with the urban environment and the social impact of transport have forced planners to broaden the scope of their analysis. Established techniques exist for the measurement and prediction of noise and considerable progress is being made in developing appropriate methods to analyse air pollution, visual intrusion and vibration. Accessibility measures are useful in examining community needs with respect to access to urban facilities (Black, 1977; Hillman *et al.*, n.d.).

The transport studies described in Part Two of this book reflect the gradual changing emphasis of transport planning during the last two decades away from long-term towards shorter-term planning horizons. In light of the general criticisms of urban transport planning mentioned so far, the following observations on transport planning practice in Canberra are made.

This need to look alternatives and to leave as many future options open as possible is crucial when the uncertainty of future events is accepted. For instance, in the early 1960s the planners thought that the level of motor cars, station wagons and other private vehicles would reach 0.45 vehicles *per capita* by the end of the century, but this level was attained by September 1976 (ABS, 1976, Table 2, p. 5) and car ownership is still rising. Long-term population and labour-force projections for Canberra are constantly under review (NCDC, 1976b). Population estimates have recently been revised downwards from 284,000 to 265,000 by June 1986 to reflect the decline in net gains from migration and a local economic recession. The uncertainties of land-use projections confirm that sensitivity analyses should become an essential part of transport planning methodology.

Since the early 1970s the need for public participation and the distractions of inquiries by parliamentary committees has increased considerably. As noted by Harrison (1978, p. 117), 'The Commission's efforts directed to advocacy and explanation cannot have been made without detracting from its planning and development operations. No authority has been required to divert so much time and energy to defending its operations.' For example, the Molonglo Arterial—which is part of the proposed freeway system—had been shown in all exhibited plans and reports for ten years or more, but strong public protests followed the announcement in 1973 that the road was to be built. Work started on the road in February 1977.

There is a continuing desire on the part of the Canberra community for effective participation in the planning process. To inform the public on particular planning proposals, and to facilitate their ability to comment and influence latter decisions, the NCDC mounts exhibitions in convenient locations throughout Canberra. The 'Plans System' comprises (a) Structure Plans accompanied by a written statement and supportive diagrams to explain the underlying principle of strategic plans and (b) Development Plans accompanied by a written statement describing the nature of the intended development, the funds involved and the time frame in which the development will take place.

Economic and environmental evaluation procedures have been an unsatisfactory part of the methodology. The two major transport studies contained no formal evaluation of alternative plans; area traffic control schemes, bus priority measures and different public transport technologies have been subjected to economic evaluations although environmental factors have largely been ignored. To overcome such deficiencies, the NCDC has recently sponsored the Canberra Air Quality Study to monitor existing air pollution and to examine meteorological and emission sources including motor vehicles. An air quality model is being developed that will predict air quality as a result of various land-use and transport planning decisions, particularly those of a broad strategic nature (NCDC, 1978a, p. 27).

In Canberra the positive achievement, emulated in relatively few cities throughout the world, is the effective co-ordination and integration between land-use planning and transport planning. Canberra is planned as a linear metropolis, which in theory can generate new town centres as it grows without overloading or congesting any of its existing centres (Stretton, 1975, p. 62). Land corridors required for the peripheral freeways to serve these free-standing 'towns' are a minimum of 500 m wide, free of any intensive urban development to allow the roadways to be fitted as naturally as possible into the landscape. Major workplaces, shopping centres, institutions, such as research establishments, colleges and hospitals, are located along the spine connecting town centres to form a land-use pattern that is supportive of line-haul public transport. If private motoring is ever forcibly reduced by high fuel prices or shortages, the Canberra community will inherit a city structure which allows the most efficient public transport system to operate amongst the major metropolitan activities and the shortest walking and cycling routes to a basic range of community facilities within each residential neighbourhood.

This book has argued that any analytical approach towards urban transport should partition the problem by geographical scale and by time horizon, and the case studies of various practical aspects of transport planning in Canberra support this proposition. Nevertheless, in practice there are real organisation problems that need to be solved to reconcile the different and sometimes incompatible aims of long-term and short-term planning (Edwards and Beimborn, 1978). The view has been expressed that there is a risk that transport system management 'may be overemphasised as an alternative rather than as a complement to capital investment in certain kinds of major transport facilities: it may only be first-aid for something that requires major surgery' (Hensher, 1979, p. 42).

Perceptions of urban transport problems and ways of solving them have altered over time and are likely to change in the future. Prior to the 1950s, public transport carried the bulk of passenger journeys and little forward planning was necessary. In the 1950s and 1960s, increased affluence, rising car ownership and traffic congestion forced highway authorities into making plans for extensive improvements to urban roads. The 1970s were characterised by a more balanced approach to co-ordinate competing transport modes, and a need to make a more efficient use of the already substantial investment in transport infrastructure. The 1980s will undoubtedly present new challenges, not least the problem of allocating scarce resources in a fair way. Irrespective of the issues defined, future transport policy will be based on more sophisticated techniques of data collection and analysis, better information, and a truer appreciation of the wide-ranging implications of alternative courses of action. The systems approach should prove to be adaptable to this new challenge.

REFERENCES

ABS. 1976. *Motor Vehicle Census—Australian Capital Territory 30 September 1976.* Canberra: Australian Bureau of Statistics

Akcelik, R. 1978. A New Look at Davidson's Travel Time Function. *Traffic Engineering and Control,* **19**, 459-63

Andrews, R.B. 1953. Mechanics of Urban Economic Base: The Problem of Terminology. *Land Economics,* **29**, 263-8

Armitage, G.A. 1977. Traffic Restraint: A Function of Parking Control? *Proceedings of the Seminar on Traffic and Environmental Management,* pp. 195-217. London: PTRC

ARRB. 1976. *Streets not Roads.* Vermont, Victoria: Australian Road Research Board

Ashford, N. and Covault, D.O. 1969. The Mathematical Form of Travel Time Factors. *Highway Research Record,* **283**, 30-47

—— and Holloway, F.M. 1972. Time Stability of Zonal Trip Production Models. *Transportation Engineering Journal of ASCE, Proceedings of the American Society of Civil Engineers,* **98**, TE 4, 799-806

Baerwald, J.E. (ed.) 1976. *Transportation and Traffic Engineering Handbook.* Englewood Cliffs, New Jersey: Prentice-Hall

Bates, J., Gunn, H. and Roberts, M. 1978. A Model of Household Car Ownership: Part 1. *Traffic Engineering and Control,* **19**, 486-91

Batty, M. 1971. Exploratory Calibration of a Retail Location Model Using Search by Golden Section. *Environment and Planning,* **3**, 411-32

—— 1975. *Urban Models—Predictions, Algorithms, Computations.* Cambridge: Cambridge University Press

—— and Mackie, S. 1972. The Calibration of Gravity, Entropy, and Related Models of Spatial Interaction. *Environment and Planning,* **4**, 131-250

—— and Sammons, R. 1978. On Searching for the Most Informative Spatial Pattern. *Environment and Planning A,* **10**, 747-79

Baxter, R.S. 1976. *Computer and Statistical Techniques for Planners.* London: Methuen

Bayliss, D. 1969. The Use of Information from Transportation Studies in Developing Parking Policies. *Proceedings of PTRC Symposium: Transportation in Very Big Cities,* pp. 21-6. London: PTRC

–– and May, A.D. 1975. The Restraint of Vehicular Traffic. Getting the Most from Our Transport Facilities: The Role of Traffic Engineering. *Transport and Road Research Laboratory Seminar.* Crowthorne

Beesley, M.E. 1973. *Urban Transport: Studies in Economic Policy.* London: Butterworths

Bell, D.A. and Symons, J.S.V. 1977. Direct Demand and Disaggregate Model Estimation. In Wigan (ed.) (1977), pp. 193-213

Ben Bouanah, J. and Stein, M.M. 1978. Urban Transportation Models: A Generalized Process for International Application. *Traffic Quarterly,* **32**, 449-70

Benjamin, B. 1968. *Demographic Analysis, Studies in Sociology – 3.* London: George Allen and Unwin

Berlinski, D. 1976. *On Systems Analysis: An Essay Concerning the Limitations of Some Mathematical Methods in the Social and Biological Sciences.* Cambridge, Massachusetts: MIT Press

Black, I.G., Gillie, R., Henderson, R. and Thomas, T. 1975. *Advanced Urban Transport.* Teakfield, Hampshire: Saxon House

Black, J. 1975. A Review of the Technical Aspects of Australian Urban Land Use/Transportation Studies. In Webb and McMaster (eds.) (1975), pp. 307-36

–– 1977. *Public Inconvenience: Access and Travel in Seven Sydney Suburbs.* Canberra: Urban Research Unit, The Australian National University

–– and Blunden, W.R. 1977. Mathematical Programming Constraints in Stategic Land Use/Transport Planning. In Sasaki and Yamaoka (eds.) (1977), pp. 649-71

–– and Conroy, M. 1977. Accessibility Measures and the Social Evaluation of Urban Structure. *Environment and Planning A,* **9**, 1013-31

–– and Salter, R.J. 1974. The Search for Generalised Coefficients in Multiple Linear Regression Equations for Trip Production from Residential Land Uses. Universities Transport Study Group Conference. Manchester

–– and Salter, R.J. 1975a. The Modelling Achievements of British Urban Land-use/Transport Studies Outside the Conurbations. *Journal of the Institution of Municipal Engineers,* **102**, 100-5

–– and Salter, R.J. 1975b. A Statistical Evaluation of the Accuracy of a Family of Gravity Models. *Proceedings of the Institution of Civil Engineers,* **59**, 1-20

Blunden, W.R. 1966. On the Traffic Effects of Frontage Land Uses on Urban Main Roads and Arterials. *Proceedings, Australian Road Research Board,* **3**, part 1, 1-18

—— 1971. *The Land Use/Transport System: Analysis and Synthesis.* Oxford: Pergamon

Bonsall, P. 1976. Tree Building with Complex Cost Structures—A New Algorithm for Incorporation into Transport Demand Models. *Transportation,* **5**, 309–29

Boyce, D.E., Day, N.D. and McDonald, C. 1969. *Metropolitan Plan Evaluation Methodology.* Philadelphia: University of Pennsylvania, Institute for Environmental Studies

BPR. 1960. US Bureau of Public Roads—Electronic Computer Program Library. *Memorandum No. 8.* Washington, DC: Bureau of Public Roads

—— 1964. *Traffic Assignment Manual.* Washington, DC: Bureau of Public Roads

—— 1965. *Calibrating and Testing a Gravity Model for any Sized Urban Area.* Washington, DC: Bureau of Public Roads

—— 1967. *Guidelines for Trip Generation Analysis.* Washington, DC: Bureau of Public Roads

—— 1970. *Urban Transportation Planning: General Information and Introduction to System 360.* Washington, DC: Bureau of Public Roads

Brambilla, R. and Longo, G. 1977. *For Pedestrians Only: Planning Design and Management of Traffic-Free Zones.* New York: Billboard Publications

Brand, D. 1974. Separable Versus Simultaneous Travel-Choice Behavior. *Transportation Research Board Special Report,* **149**, 187–206

Brennan, F. 1971. *Canberra in Crisis: A History of Land Tenure and Leasehold Administration.* Canberra: Dalton Publishing Company

Brierley, J. 1972. *Parking of Motor Vehicles,* 2nd edn. London: Applied Science Publishers

Brown, A.J. and Sherrard, H.M. 1951. *Town and Country Planning.* Melbourne: Melbourne University Press

Bruton, M.J. 1975. *Introduction to Transportation Planning,* 2nd edn. London: Hutchinson

BTE. 1972. *Economic Evaluation of Capital Investment in Urban Public Transport.* Canberra: Government Printers

Buchanan, C.D. 1963. *Traffic in Towns: A Study of the Long Term Problems of Traffic in Urban Areas—Report of the Working Group.* London: HMSO

Burrage, R.H. and Mogren, E.G. 1957. *Parking.* Saugatuck, Connecticut: Eno Foundation for Highway Traffic Control

Burrell, J.E. 1969. Multiple Route Assignment and its Application to Capacity Restraint. In Leutzbach and Baron (eds.) (1969), pp. 210–19

Caldwell, J.R. 1979. Low Cost Public Transport Improvements. *Fifth Australian Transport Research Forum Papers,* pp. 690-704

Carrothers, G.A.P. 1956. An Historical Review of the Gravity and Potential Concepts of Human Interaction. *Journal of the American Institute of Planners,* **22,** 94-102

CATS. 1962. *Canberra Area Transportation Study Report on 1961 Surveys Prepared for the National Capital Development Commission, Canberra A.C.T.* Canberra: Rankine and Hill and De Leuw, Cather and Company

—— 1963. *Canberra Area Transportation Study Engineering Report for the National Capital Development Commission.* Canberra: Rankine and Hill and De Leuw, Cather and Company

Chadwick, G. 1978. *A Systems View of Planning: Towards a Theory of Urban and Regional Planning Process,* 2nd edn. Oxford: Pergamon

Chapin, F.S., Jr. 1976. *Urban Land-Use Planning,* 2nd edn. Urbana, Illinois: University of Illinois Press

Clark, N. 1975. Urban Public Transport in Australia. In Webb and McMaster (eds.) (1975), pp. 450-69

Clark, Nicholas and Associates, 1978. *A Preliminary Investigation into the Impact of Boom Gates on City Parking, Canberra.* Canberra: Nicholas Clark and Associates

Cleveland, D.E. 1976. Traffic Studies. In Baerwald (ed.) (1976), pp. 404-70

Cliff, A.D., Haggett, P., Ord, J.K., Bassett, K. and Davies, R. 1975. *Elements of Spatial Structure.* Cambridge: Cambridge University Press

Collins, M.F. and Pharoah, T.M. 1974. *Transport Organisation in a Great City: The Case of London.* London: George Allen and Unwin

Conroy, M.M. 1978. Accessibility and the Evaluation of Land Use/ Transportation Plans. 9th Australian Road Research Board Conference, Session 25, pp. 1-17

Cox, P.R. 1970. *Demography,* 4th edn. Cambridge: Cambridge University Press

Creighton, Hamburg Planning Consultants. 1971. Data Requirements for Metropolitan Transportation Planning. *National Cooperative Highway Research Program Report,* **120.** Washington, DC: Highway Research Board

Creighton, R.L. 1970. *Urban Transportation Planning.* Urbana, Illinois: University of Illinois Press

Cresswell, R. (ed.) 1977 *Passenger Transport and the Environment.* London: Leonard Hill

CSTTPS. 1976. *Canberra Short Term Transport Planning Study: Travel Surveys and Data Assembly*, vol. I. Parkside, South Australia: P.G. Pak-Poy and Associates, and John Paterson Urban Systems

—— 1977a. *Canberra Short Term Transport Planning Study: Study Report.* Parkside, South Australia: P.G. Pak-Poy and Associates in association with John Paterson Urban Systems

—— 1977b. *Canberra Short Term Transport Planning Study: Technical Report on the Development of the Transport Demand Projection Models.* Parkside, South Australia: P.G. Pak-Poy and Associates and John Paterson Urban Systems

—— 1977c. *Canberra Short Term Transport Planning Study: Technical Report on the Development of the Disaggregate Choice Models.* Parkside, South Australia: P.G. Pak-Poy and Associates in association with John Paterson Urban Systems

Daganzo, C.F. 1977a. On the Traffic Assignment Problem With Flow Dependent Costs–I. *Transportation Research*, 11, 433-7

—— 1977b. Some Statistical Problems in Connection With Traffic Assignment. *Transportation Research*, 11, 385-9

—— and Sheffi, Y. 1977. On Stochastic Models of Traffic Assignment. *Transportation Science*, 11, 253-74

Daverin, D.L. 1977. Canberra: Current Analytical Transport Planning Requirements. In Wigan (ed.) (1977), pp. 59-63

Davidson, K.B. 1966. A Flow Travel Time Relationship for Use in Transportation Planning. *Proceedings, Australian Road Research Board*, 3, part 1, 183-94

de Donnea, F.X. 1971. *The Determinants of Transport Mode Choice in Dutch Cities: Some Disaggregate Stochastic Models.* Rotterdam: Rotterdam University Press

Dial, R.B. 1971. A Probabilistic Multipath Assignment which Obviates Path Enumeration. *Transportation Research*, 5, 83-111

—— 1976. Urban Transportation Planning System: Philosophy and Function. *Transportation Research Record*, 599, 43-8

DOE. 1975. *Calculation of Road Traffic Noise.* Department of the Environment and the Welsh Office Joint Publication. London: HMSO

Domencich, T.A. and McFadden, D. 1975. *Urban Travel Demand: A Behavioral Analysis.* Amsterdam: North Holland

Dorrington, T.E. 1971. Some New Methods of Applying and Calibrating the SELNEC Modal Split Model. *Mathematical Advisory Unit, MAU Note*, 233. London: Ministry of Transport

DOT. 1974. *Software Systems Development Program–Introduction to*

Urban Travel Demand Forecasting, Volume 1. Washington DC: Department of Transportation

Doubleday, C. 1977. Some Studies of the Temporal Stability of Person Trip Generation Models. *Transportation Research,* 11, 255–63

Downes, J.D. and Gyenes, L. 1976. Temporal Stability and Forecasting Ability of Trip Generation Models in Reading. *TRRL Report,* 726. Crowthorne, Berkshire: Transport and Road Research Laboratory

DpT. 1977. *Residential Roads and Footpaths.* London: Department of Transport

Drake, J.W. 1973. *The Administration of Transportation Modeling Projects.* Lexington, Massachusetts: Lexington Books

DT. 1978. *Road Safety Guidelines for Town Planning.* Canberra: Australian Government Printing Service

Duhs, L.A. and Beggs, J.J. 1977. The Urban Transportation Study. In Hensher (ed.) (1977b), pp. 228–51

Edwards, J.L. and Beimborn, E.A. 1978. Future Relationships Between Long- and Short-Range Urban Transportation Planning. *Traffic Quarterly,* 32, 531–44

Elkington, J., McGlynn, R. and Roberts, J. 1977. *The Pedestrian: Planning and Research, A Literature Review and Annotated Bibliography.* London: Transport and Environment Studies

Elliot, J.R. 1977. Controlled Parking in Residential Areas. *Proceedings of the Seminar on Traffic and Environment Management,* pp. 240–9 London: PTRC

Ellis, R.H. 1977. Parking Management Strategies. *Transportation Research Board Special Report,* 172, 55–66

Evans, A.W. 1970. Some Properties of Trip Distribution Models. *Transportation Research,* 4, 19–36

—— 1971. The Calibration of Trip Distribution Models with Exponential or Similar Functions. *Transportation Research,* 5, 15–38

Ferland, J.A., Florian, M. and Achim, C. 1975. On Incremental Methods for Traffic Assignment. *Transportation Research,* 9, 237–9

Fertal, M.J., Weiner, E., Balek, A.J. and Sevin, A.F. 1970. *Modal Split— Documentation of Nine Methods for Estimating Transit Usage.* Washington, DC: Department of Transportation

Field, J.F., Hallam, C.E. and Colston, M. 1979. The Impact of Land Use on Traffic. *Fifth Australian Transport Research Forum Papers,* pp. 184–97

Fisher, G.P. (ed.) 1978. *Goods Transportation in Urban Areas, Proceedings of the Engineering Foundation Conference, Sea Island, Georgia, December 4–9, 1977.* Washington, DC: Department of Transportation

Florian, M. and Fox, B. 1976. On the Probabilistic Origin of Dial's Multipath Traffic Assignment Model. *Transportation Research,* **10**, 339-41

— and Nguyen, S. 1976. An Application and Validation of Equilibrium Trip Assignment Method. *Transportation Science,* **10**, 374-90

Fowkes, A.S. and Button, K.J. 1977. An Evaluation of Car Ownership Forecasting Techniques. *International Journal of Transport Economics,* **4**, 115-43

Fruin, J.J. 1971. *Pedestrian Planning and Design.* New York: Metropolitan Association of Urban Designers and Environmental Planners

Fullerton, B. 1975. *The Development of British Transport Networks.* London: Oxford University Press

Gakenheimer, R. and Meyer, M. 1978. *Transportation System Management The Record and a Look Ahead–Summary Report.* Washington, DC: Department of Transportation

Garner, D. 1974. Transit Planning and Characteristics of Ontario Transit Operations. In Shortreed (ed.) (1974), pp. 1-38

Gibberd, F. 1962. *Town Design,* 4th edn. London: the Architectural Press

GLC. 1976. *Freight in London: Freight Policy Background Information.* London: Greater London Council/London Freight Conference

Golding, S. 1972. A Category Analysis Approach to Trip Generation. *Proceedings, Australian Road Research Board,* **6**, part 2, 306-24

Golob, T.F., Canty, E.T., Gustafson, R.L. and Vitt, J.E. 1972. An Analysis of Consumer Preferences for a Public Transportation System. *Transportation Research,* **6**, 81-102

—, Horowitz, A.D. and Wachs, M. 1979. Attitude-Behaviour Relationships in Travel Demand Modeling. In Hensher and Stopher (eds.) (1979), pp. 739-57

Goodwin, P.B. 1978. On Grey's Critique of Generalised Costs. *Transportation,* **7**, 281-95

Hall, A.D. and Fagen, R.E. 1956. Definition of System. *General Systems,* **1**, 18-28

Hamer, A.M. 1976. *The Selling of Rail Rapid Transit: A Critical Look at Urban Transportation Planning.* Lexington, Massachusetts: Lexington Books

Hansen, W.G. 1959. How Accessibility Shapes Land Use. *Journal of the American Institute of Planners,* **25**, 73-6

— 1962. Evaluation of Gravity Model Trip Distribution Procedures. *Highway Research Board Bulletin,* **347**, 67-76

Harrison, A.J. 1974. *The Economics of Transport Appraisal.* New York: Halsted Press

Harrison, P. 1964. Long Term Planning for Canberra. *Australian Planning Institute Journal*, 2, 263-6

—— 1978. Australian Capital Territory. In Ryan (ed.) (1978), pp. 109-19

Hartgen, D.T. and Tanner, G.H. 1971. Investigations of the Effect of Traveler Attitudes in a Model of Mode-Choice Behavior. *Highway Research Record*, **369**, 1-14

Hasell, B.B., Foulkes, M. and Robertson, J.J.S. 1978. Freight Planning in London: Reducing the Environmental Impact. *Traffic Engineering and Control*, **19**, 182-5

Hayfield, C.P. and Stoker, R.B. 1978. The Geographical Stability of a Typical Trip Production Model. Applications of National and Local Data in Four Urban Areas. *Transportation*, 7, 211-24

Healey, P. 1977. The Sociology of Urban Transportation Planning: A Socio-Political Perspective. In Hensher (ed.) (1977b), pp. 199-227

Hemmens, G.C. 1968. Survey of Planning Agency Experience with Urban Development Models, Data Processing and Computers. *Highway Research Board, Special Report*, **97**, 219-30

Hensher, D.A. 1972. The Consumer's Choice Function: A Study of Traveller Behaviour and Values. Unpublished PhD thesis, University of New South Wales

—— 1977a. *Value of Business Travel Time.* Oxford: Pergamon

—— (ed.) 1977b. *Urban Transport Economics.* Cambridge: Cambridge University Press

—— 1977c. Demand for Urban Passenger Transport. In Hensher (ed.) (1977b), pp. 72-99

—— 1979. Urban Transport Planning: The Changing Emphasis. *Search: The Journal of ANZAAS*, **10**, 42-8

—— and Stopher, P.R. (eds.) 1979. *Behavioural Travel Modelling.* London: Croom Helm

Hill, D.M. and Dodd, N. 1966. Studies of Trends of Travel Between 1954 and 1964 in a Large Metropolitan Area. *Highway Research Record*, **141**, 1-23

—— and Von Cube, H.G. 1963. Development of a Model for Forecasting Travel Mode Choice in Urban Areas. *Highway Research Record*, **38**, 78-96

Hillman, M., Henderson, I. and Whalley, A. n.d. *Transport Realities and Planning Policy: Studies of Friction and Freedom in Daily Travel.* London: Political and Economic Planning, no. 567

—— and Whalley, A. 1975. Land Use and Travel. *Built Environment*, **1**, 105-11

Hirsch, W.Z. 1973. *Urban Economic Analysis.* New York: McGraw-Hill

Hoel, P.G. 1962. *Introduction to Mathematical Statistics,* 3rd edn. New York: John Wiley

Holford, Sir William. 1972. Tomorrow's Canberra: A Review Article. *Town Planning Review,* **43**, 26-30

Holland, E.P. and Watson, P.L. 1978. Traffic Restraint in Singapore. *Traffic Engineering and Control,* **19**, 14-22

Holroyd, J. and Hillier, J.A. 1969. Area Traffic Control in Glasgow: A Summary of Results from Four Control Schemes. *Traffic Engineering and Control,* **11**, 220-3

–– 1971. Further Results of Area Traffic Control in Glasgow. *Traffic Engineering and Control,* **13**, 195-8

Holton, J.A. and Pattinson, W.H. 1976. Survey of and Comments on Residential Street Planning and Standards. In ARRB (1976), pp. 30-40

Hotchkiss, W.E. 1977. Cost-Benefit Analysis. In Hensher (ed.) (1977b), pp. 44-54

Hudson, M. 1978. *The Bicycle Planning Book.* London: Open Books/ Friends of the Earth

Hunt, P.B. and Kennedy, J.V. 1978. TRANSYT Version 7. Unpublished Note, Transport and Road Research Laboratory, Crowthorne, Berkshire

Hutchinson, B.G. 1974. *Principles of Urban Transport Systems Planning.* New York: McGraw-Hill

–– and Smith, D.P. 1978. The Journey-to-Work in Urban Canada. 9th Australian Road Research Board Conference. Brisbane

IPT. 1976. *Intertown Public Transport: Alternatives for Canberra,* 2nd edn. Canberra: National Capital Development Commission

ITTE. 1976. *Trip Generation: An Institute of Transportation Engineers Informational Report.* Arlington, Virginia: Institute of Transportation Engineers

Jones, I.S. 1977. *Urban Transport Appraisal.* London: Macmillan

Kassoff, H. and Deutschman, H.D. 1969. Trip Generation: A Critical Appraisal. *Highway Research Record,* **297**, 15-30

Keyani, B.I. and Putnam, E.S. 1976. *Transportation System Management: State of the Art.* Santa Barbara, California: INTERPLAN Corporation

Kirby, H.R. 1974. Theoretical Requirements for Calibrating Gravity Models. *Transportation Research,* **8**, 97-104

Lane, R., Powell, T.J. and Prestwood-Smith, P. 1971. *Analytical Transport Planning.* London: Duckworth

Lang, A.S. and Soberman, R.M. 1964. *Urban Rail Transit: Its Economics and Technology.* Cambridge, Massachusetts: MIT Press

Lee, C. 1973. *Models in Planning: An Introduction to the Use of Quantitative Models in Planning.* Oxford: Pergamon

Lee, Y. and DePhillips, F.C. 1979. Planning for Twelve Million Annual Visitors to Atlantic City. *Traffic Quarterly,* **33,** 61-81

Leibbrand, K. 1970. *Transportation and Town Planning.* London: Leonard Hill

Leutzbach, W. and Baron, P. (eds.) 1969. *Beiträge zur Theorie das Verkehrsflusses: Proceedings, 4th International Symposium on the Theory of Traffic Flow.* Bonn: Herausgegeben vom Bundesminister für Verkehr

Leveris, B. 1975. Australian Transportation Studies: An Appraisal Using the Delphi Approach. *Australian Road Research Board, Internal Report,* 2. Vermont, Victoria: Australian Road Research Board

Levin, M.R. and Abend, N.A. 1971. *Bureaucrats in Collision: Case Studies in Area Transportation Planning.* Cambridge, Massachusetts: MIT Press

Levinson, H.S. 1978. Characteristics of Urban Transportation Demand —A New Data Bank. Paper presented at the 57th Annual Meeting of the Transportation Research Board, Washington, DC

Lichfield, N. 1966. Cost Benefit Analysis in Town Planning—A Case Study: Swanley. *Urban Studies,* **3,** 215-49

Linge, G.J.R. 1975. *Canberra Site and City.* Canberra: Australian National University Press

Lockwood, S.C. and Wagner, F.A. 1977. Methodological Framework for the TSM Planning Process. *Transportation Research Board Special Report,* **172,** 100-18

Lowry, I.S. 1964. *A Model of Metropolis.* Santa Monica, California: RAND Corporation

Lynch, K. 1971. *Site Planning,* 2nd edn. Cambridge, Massachusetts: MIT Press

McFadden, D. and Reid, F. 1974. Aggregate Travel Demand Forecasting From Disaggregated Behavioral Models. *Transportation Research Board,* **534,** 24-37

McIntosh, P.T. and Quarmby, D.A. 1970. Generalised Costs and the Estimation of Movement Costs and Benefits in Transport Planning. *Mathematical Advisory Unit, MAU Note,* 179. London: Department of the Environment

Mackinder, I.H. 1972. Compact: A Simple Transportation Planning Package. *Traffic Engineering and Control,* **13,** 512-16

——, Evans, S.E. and May, R. 1975. The Distribution of Household Income in Trip Generation. *Traffic Engineering and Control,* **16**, pp. 546-8, p. 556

McLoughlin, J.B. 1969. *Urban and Regional Planning: A Systems Approach.* London: Faber and Faber

Martin, B.V., Memmott III, F.W. and Bone, A.J. 1961. *Principles and Techniques of Predicting Future Demand for Urban Area Transportation.* Cambridge, Massachusetts: MIT Press

Mason, P.R. 1972. Multiflow Multiple Routing Procedure. *Traffic Engineering and Control,* **14**, 280-2

Matson, T.M., Smith, W.S. and Hurd, F.W. 1955. *Traffic Engineering.* New York: McGraw-Hill

May, A.D. and Westland, D. 1979. Transportation System Management: TSM-Type Projects in Six Selected European Countries. *Traffic Engineering and Control Supplement.* London: Printerhall

Menon, A.P.G., Tierney, A.J.H., Blunden, W.R. and Tindall, J.I. 1974. A Study of Davidson's Flow Travel Time Relationship. *Proceedings, Australian Road Research Board,* **7**, 5-20

Miller, A.J. 1968. The Capacity of Signalised Intersections in Australia. *Australian Road Research Board, Bulletin, 3.* Vermont, Victoria: Australian Road Research Board

Mishan, E.J. 1971. *Cost-Benefit Analysis.* London: George Allen and Unwin

Morris, R.L. and Zisman, S.B. 1962. The Pedestrian, Downtown and the Planner. *Journal of the American Institute of Planners,* **28**, 152-8

Mullen, P. and White, M. 1977. Forecasting Car Ownership: A New Approach—Part 2. *Traffic Engineering and Control,* **18**, 422-6

Musgrave, R.A. and Musgrave, P.B. 1973. *Public Finance in Theory and Practice.* New York: McGraw-Hill

Nairn, R.J. and Partners. 1978. *Case Studies in Land Use/Transport Interaction Using 'TRANSTEP'.* Canberra: R.J. Nairn and Partners

—— 1979. *City Area Traffic Control: Traffic Operations.* Canberra: R.J. Nairn and Partners

NATO. 1976. *Bus Priority Systems: NATO Committee on the Challenges of Modern Society. CCMS Report No. 45.* Crowthorne, Berkshire: Transport and Road Research Laboratory

Navin, F.P.D. and Schultz, G.W. 1970. A Technique to Calibrate Choice Models. *Highway Research Record,* **322**, 68-76

NCDC. 1970. *Tomorrow's Canberra.* Canberra: Australian National University Press

—— 1975. Planning Brief for Lanyon Territorial Unit. *NCDC Technical Paper*, 11. Canberra: National Capital Development Commission

—— 1976a. Metropolitan Structure Plan Review. National Capital Development Commission Internal Working Document only

—— 1976b. Population Projections for Canberra 1976-1986. *NCDC Technical Paper*, 12. Canberra: National Capital Development Commission

—— 1977. Evaluation of Area Traffic Control Measures for the City Centre. *Report of an In-House Technical Study for the National Capital Development Commission.* Canberra: R.J. Nairn and Partners

—— 1978a. *National Capital Development Commission 21st Annual Report.* Canberra: Wilkie

—— 1978b. Statement of Engineering Planning Objectives and Criteria for City Area Traffic Control Stage I (1978/79) National Capital Development Commission File No. 75/833

—— 1978c. TRANSYT Version 6N. *National Capital Development Commission Engineering, Internal Report*, No. 2/78

—— 1978d. Parking Accumulation and Inventories in Canberra 1975-1978: Main Report. *National Capital Development Commission Engineering, Internal Report*, No. 4/78

Neuburger, H. 1971. User Benefit in the Evaluation of Transport and Land Use Plans. *Journal of Transport Economics and Policy, 5,* 52-75

Neutze, M. 1977. *Urban Development in Australia–A Descriptive Analysis.* Sydney: George Allen and Unwin

—— 1978. *Australian Urban Policy.* Sydney: George Allen and Unwin

Newell, G.F. and Vuchic, V.R. 1968. Rapid Transit Interstation Spacings for Minimum Travel Time. *Transportation Science*, 2, 303-9

OECD. 1974. *Urban Traffic Models: Possibilities for Simplification.* Paris: Organisation for Economic Co-operation and Development

—— 1977. *Integrated Urban Traffic Management.* Paris: Organisation for Economic Co-operation and Development

O'Flaherty, C.A. 1974. *Highways Vol. 1–Highways and Traffic,* 2nd edn. London: Edward Arnold

—— and Parkinson, M.H. 1972. Movement on a City Centre Footway. *Traffic Engineering and Control*, 13, 434-8

Openshaw, S. 1978. An Empirical Study of Some Zone-Design Criteria. *Environment and Planning A*, 10, 781-94

—— and Connolly, C.J. 1977. Empirically Derived Deterrence Functions for Maximum Performance Spatial Interaction Models. *Environment and Planning A*, 9, 1067-79

Pak-Poy, P.G. and Associates. 1967. *Joint Study of Public Transport for Canberra.* Parkside, South Australia: P.G. Pak-Poy and Associates
—— 1973. *Bus Priority Scheme–Woden/City/Belconnen–Feasibility Study.* Parkside, South Australia: P.G. Pak-Poy and Associates

Papoulias, D.B. and Dix, M.C. 1978. Results of Surveys in Oxford to Investigate the Impact of Bus Lane Schemes. *Traffic Engineering and Control,* **19**, 26-31, 42

Paterson, John, *et al.* 1974. *Canberra Modal Split Study.* North Melbourne, Victoria: John Paterson Urban Systems

Pierce, J.R. and Wood, K. 1977. Bus TRANSYT–A User's Guide. *TRRL Supplementary Report,* 266. Crowthorne, Berkshire: Transport and Road Research Laboratory

Pitfield, D.E. 1978. Algorithm 6: The χ^2 Test for Predicted Trip Matrices. *Environment and Planning A,* **10**, 1201-6

Plowden, S.P.C. 1967. Transportation Studies Examined. *Journal of Transport Economics and Policy,* **1**, 5-27

Quinby, H.D. 1976. Mass Transportation Characteristics. In Baerwald (ed.) (1976), pp. 207-57

Rallis, T. 1967. Capacity of Transport Centres: Ports, Railway Stations, Road Haulage Centres and Airports. *Department for Road Construction, Transportation Engineering and Town Planning, Report 35.* Copenhagen: Technical University of Denmark

Ratcliffe, E.P. 1972. A Comparison of Driver's Route Choice Criteria and Those Used in Current Assignment Processes. *Traffic Engineering and Control,* **13**, 526-9

Recker, W.W. and Golob, T.F. 1976. An Attitudinal Modal Choice Model. *Transportation Research,* **10**, 299-310

Rees, P.H., Smith, A.P. and King, J.R. 1977. Population Models. In Wilson *et al.* (eds.) (1977), pp. 49-129
—— and Wilson A.G. 1977. *Spatial Population Analysis.* London: Edward Arnold

Rice, P. 1977. Practical Urban Railway Capacity–A World Review. In Sasaki and Yamaoka (eds.) (1977), pp. 773-808

Rigby, J.P. 1977. An Analysis of Travel Patterns Using the 1972/73 National Travel Survey. *TRRL Report* 790. Crowthorne, Berkshire: Transport and Road Research Laboratory

Robbins, J. 1978. Mathematical Modelling–The Error of Our Ways. *Traffic Engineering and Control,* **19**, 32-5

Robertson, D.I. 1969a. TRANSYT: A Traffic Network Study Tool. *Road Research Laboratory Report* 253. Crowthorne, Berkshire: Road Research Laboratory

—— 1969b. TRANSYT Method for Area Traffic Control. *Traffic Engineering and Control*, **11**, 276-81

Robillard, P. 1975. Estimating O-D Matrix from Observed Link Volumes. *Transportation Research*, **9**, 123-8

Robinson, C.C. 1976. Highway Capacity. In Baerwald (ed.) (1976), pp. 309-76

Rogers, K.G., Townsend, G. and Metcalf, A.E. 1971. Planning for the Work Journey—A Generalised Explanation of Modal Choice. *Local Government Operational Research Unit,* Report, C.67

RRL. 1965. *Research on Road Traffic.* London: HMSO

Ruiter, E.R. 1967. Improvements in Understanding, Calibrating, and Applying the Opportunity Model. *Highway Research Record*, **165**, 1-21

Ryan, P.F. (ed.) 1978. *Urban Management Processes.* Canberra: Australian Government Publishing Service

Sandercock, L. 1975. *Cities for Sale: Property, Politics and Urban Planning in Australia.* Melbourne: Melbourne University Press

Sasaki, T. and Yamaoka, T. (eds.) 1977. *Proceedings of the Seventh International Symposium on Transportation and Traffic Theory.* Kyoto: Institute of System Science Research

Sayer, R.A. 1976. A Critique of Urban Modelling: From Regional Science to Urban and Regional Political Economy. *Progress in Planning*, **6**, 187-254

Schocken, T.D. 1968. Splitting Headaches. *Traffic Quarterly*, **22**, 289-96

Schumacher, E.F. 1978. *A Guide for the Perplexed.* London: Sphere Books

Schofer, J.L. and Stopher, P.R. 1979. Specifications for a New Long-Range Urban Transportation Planning Process. *Transportation*, **8**, 199-218

Seddon, P.A. 1972. Another Look at Platoon Dispersion—3. The Recurrence Relationship. *Traffic Engineering and Control*, **13**, 442-4

Selinger, C.S. 1977. Managing Transportation Demand by Alternative Work Schedule Techniques. *Transportation Research Board Special Report*, **172**, 67-73

Senior, M.L. and Williams, H.C.W.L. 1977. Model-Based Transport Policy Assessment—1: The Use of Alternative Forecasting Models. *Traffic Engineering and Control*, **18**, 402-6

Sharp, C. and Jennings, T. 1976. *Transport and the Environment.* Leicester: Leicester University Press

Shortreed, J.H. (ed.) 1974. *Urban Bus Transit—A Planning Guide.* Waterloo, Ontario: Department of Civil Engineering, University of Waterloo

Shoup, D.C. and Pickrell, D.H. 1978. Problems with Requirements in Zoning Ordinances. *Traffic Quarterly,* **32**, 545-61

Sims, A.G. 1979. The Sydney Co-Ordinated Adaptive Traffic System. Research Directions in Computer Control of Urban Traffic Systems, Engineering Foundation Conference. Pacific Grove, California

Smith, T.B. and Leigh, C.M. 1977. Regional Economic Models. In Wilson *et al.* (eds.) (1977), pp. 131-207

Solesbury, W. and Townsend, A. 1970. Transportation Studies and British Planning Practice. *Town Planning Review,* **41**, 63-79

Sosslau, A.B., Hassam, A.B., Carter, M.M. and Wickstrom, G.V. 1978. Quick-Response Urban Travel Estimation Techniques and Transferable Parameters—User's Guide. *National Cooperative Highway Research Program Report,* **187**, Washington, DC: Transportation Research Board

——, Heanue, K.E. and Balek, A.J. 1964. Evaluation of a New Modal Split Procedure. *Public Roads,* **33**, 5-19

Spear, B.D. 1976. Generalized Attribute Variable for Models of Mode Choice Behavior. *Transportation Research Board,* **592**, 6-11

Stanford Research Institute. 1978. Quantifying the Benefits of Separating Pedestrians and Vehicles. *National Cooperative Highway Research Program Report,* **189.** Washington, DC: Transportation Research Board

Starkie, D.N.M. 1976. *Transportation Planning, Policy and Analysis.* Oxford: Pergamon

Stopher, P.R. and Meyburg, A.H. 1975. *Urban Transportation Modeling and Planning.* Lexington, Massachusetts: Lexington Books

—— and Meyburg, A.H. 1976. *Transportation Systems Evaluation.* Lexington, Massachusetts: Lexington Books

Stretton, H. 1975. *Ideas for Australian Cities,* 2nd edn. Melbourne: Georgian House

Sweet, C.E., Jr. 1969. *Guidelines for the Administration of Urban Transportation Planning.* Washington, DC: Institute of Traffic Engineers

Symons, J.S.V. and Bell, D.A. 1977. POLSET: Policy Analysis Using Disaggregate Models. In Wigan (ed.) (1977), pp. 117-25

Taaffe, E.J. and Gauthier, H.L., Jr. 1973. *Geography of Transportation.* Englewood Cliffs, New Jersey: Prentice-Hall

Tagliacozzo, F. and Pirzio, F. 1973. Assignment Models and Urban

Path Selection Criteria: Results of a Survey of the Behaviour of Road Users. *Transportation Research,* 7, 313-29

Talvitie, A. and Kirshner, D. 1978. Specification, Transferability and Effect of Data Outliers in Modeling the Choice of Mode in Urban Travel. *Transportation,* 7, 311-31

Tanner, J.C. 1962. Forecasts of Future Numbers of Vehicles in Great Britain. *Roads and Road Construction,* 40, 263-74

–– 1974. Forecasts of Vehicles and Traffic in Great Britain. *TRRL Report,* 650. Crowthorne, Berkshire: Transport and Road Research Laboratory

–– 1977. Car Ownership Trends and Forecasts. *TRRL Report,* 799. Crowthorne, Berkshire: Transport and Road Research Laboratory

––, Gyenes, L., Lynam, D.A., Magee, S. and Tulpule, A.H. 1973. The Development and Calibration of the CRISTAL Transport Planning Model. *TRRL Report,* 574. Crowthorne, Berkshire: Transport and Road Research Laboratory

TAU. 1977. *Some Effects of Flexible Working Hours on Traffic Conditions at a Large Office Complex.* London: Traffic Advisory Unit, Department of Transport

Taylor, M.A.P. 1977. Parameter Estimation and Sensitivity of Parameter Values in a Flow-Rate/Travel-Time Relation. *Transportation Science,* 11, 275-92

Thomson, J.M. 1969. *Motorways in London.* London: Duckworth

–– 1977. *Great Cities and Their Traffic.* London: Victor Gollancz

Tranter, C.J. 1957. *Techniques of Mathematical Analysis.* London: English Universities Press

Tripp, Sir H. Alker. 1942. *Town Planning and Road Traffic.* London: Edward Arnold

Troy, P. and Neutze, M. 1969. Urban Road Planning in Theory and Practice. *Journal of Transport Economics and Policy,* 3, 139-51

Turner, E.D. and Giannopoulos, G.A. 1974. Pedestrianisation: London's Oxford Street Experiment. *Transportation,* 3, 95-126

Van Vliet, D. 1976. Road Assignment ii–The GLTS Model. *Transportation Research,* 10, 145-50

Voorhees, Alan M. and Associates. 1967. *Canberra Land Use Transportation Study: General Concept Plan.* McLean, Virginia: Alan M. Voorhees and Associates

–– 1968. Factors and Trends in Trip Lengths. *National Cooperative Highway Research Progam Report,* 48. Washington, DC: Highway Research Board

–– 1974a. *Canberra Public Transport Study–Analysis of Public Transport Alternatives.* Melbourne: Alan M. Voorhees and Associates

—— 1974b. *Canberra Public Transport Study–Technical Report No. 2: Transport Models and Data*. Melbourne: Alan M. Voorhees and Associates

Vreugdenhil, J.J. 1976. Traffic Management Schemes for Existing Residential Street Layouts on the Grid System. In ARRB (1976), pp. 1-19

Walker, J.R. 1968. Rank Classification: A Procedure for Determining Future Trip Ends. *Highway Research Record*, **240**, 88-99

Wallis, I.P. 1979. Private Bus Operations in Urban Areas–Their Economics and Role. *Fifth Australian Transport Research Forum Papers*, pp. 705-22

Wardrop, A.W. 1979. In Search of Standards of Service for Urban Public Transport. *Fifth Australian Transport Research Forum Papers*, pp. 640-62

Wardrop, J.G. 1952. Some Theoretical Aspects of Road Traffic Research. *Proceedings, Institution of Civil Engineers*, Part II, 1, 325-78

—— and Charlesworth, G. 1954. A Method of Estimating Speed and Flow of Traffic from a Moving Vehicle. *Proceedings, Institution of Civil Engineers*, Part II, 3, 158-71

Watson, G.I. 1978. Public Transport Passenger Surveys. *Traffic Engineering and Control*, **19**, 268-72

Watson, P.L. 1974. Choice of Estimation Procedure for Models of Binary Choice: Some Statistical and Empirical Evidence. *Regional and Urban Economics*, 4, 187-200

Watters, B.J. 1972. Planning Requirements for Goods Delivery Facilities. Unpublished M.Eng.Sci. thesis, University of New South Wales

Weaver, T.E. and Riley, T.D. 1971. A Note on the Convergence of the Traffic Assignment Process. *Traffic Engineering and Control*, **13**, 572-5

Webb, G.R. 1975. The Economics of Government Bus Services in Australia. In Webb and McMaster (eds.) (1975), pp. 5-16

—— and McMaster, J.C. (eds.) 1975. *Australian Transport Economics: A Reader*. Sydney: Australia and New Zealand Book Company

Webb, L.M. and Cooper, I.G. 1976. Transport Research, Planning and Operations Within the Department of the Capital Territory. *Australian Transport Research Forum: Forum Papers Adelaide 1976*, pp. 215-33

Weiner, E. 1969. Modal Split Revisited. *Traffic Quarterly*, **23**, 5-28

Wells, G.R. 1975. *Comprehensive Transport Planning*. London: Charles Griffin

White, P.R. 1976. *Planning for Public Transport*. London: Hutchinson

Wigan, M.R. (ed.) 1977. *New Techniques for Transport Systems Analysis.* Vermont, Victoria: Australian Road Research Board

Wilbur Smith and Associates. 1966. *Transportation and Parking for Tomorrow's Cities.* New Haven, Connecticut: Wilbur Smith and Associates

—— 1977. *Characteristics of Urban Transportation Demand.* New Haven, Connecticut: Wilbur Smith and Associates

Williams, H.C.W.L. 1977. On the Formulation of Travel Demand Models and Economic Evaluation Measures of User Benefit. *Environment and Planning A,* **9**, 285–344

—— and Senior, M.L. 1977. Model-Based Transport Policy Assessment –2: Removing Fundamental Inconsistencies from the Models. *Traffic Engineering and Control,* **18**, 464–9

Williams, I.N. 1977. Algorithm 1: Three-Point Rational Function Interpolation for Calibrating Gravity Models. *Environment and Planning A,* **9**, 215–21

Wills, L.H. 1976. Discussion on 'Residential Streets: Alternatives to the Conventional'. In ARRB (1976), pp. 44–6

Wilson, A.G. 1967. A Statistical Theory of Spatial Distribution Models. *Transportation Research,* **1**, 253–69

—— 1969. The Use of Entropy Maximising Methods in the Theory of Trip Distribution, Mode Split and Route Assignment. *Journal of Transport Economics and Policy,* **3**, 108–26

—— 1970. *Entropy in Urban and Regional Modelling.* London: Pion

—— 1974. *Urban and Regional Models in Geography and Planning.* New York: John Wiley

——, Hawkins, A.F., Hill, G.J. and Wagon, D.J. 1969. Calibrating and Testing the SELNEC Transport Model. *Regional Studies,* **3**, 337–50

—— and Kirkby, M.J. 1975. *Mathematics for Geographers and Planners.* Oxford: Clarendon Press

——, Rees, P.H. and Leigh, C.M. (eds.) 1977. *Models of Cities and Regions: Theoretical and Empirical Developments.* New York: John Wiley

Wilson, F.S. 1967. *Journey to Work–The Modal Split.* London: McLarens

Wilson, S.R. 1976. Statistical Notes on the Evaluation of Calibrated Gravity Models. *Transportation Research,* **10**, 343–5

Winsten, C.B. 1967. Regression Analysis Versus Category Analysis. *Seminar Proceedings,* pp. 17–19. London: PTRC

Witheford, D.K. 1976. Urban Transportation Planning. In Baerwald (ed.) (1976), pp. 502–58

Wood, A.A. 1977. Foot Streets and Public Transport. In Cresswell (ed.) (1977), pp. 95-114

Wootton, H.J. and Pick, G.W. 1967. A Model for Trips Generated by Households. *Journal of Transport Economics and Policy*, **1**, 137-53

Zupan, J.M. 1968. Mode Choice: Implications for Planning. *Highway Research Record*, **251**, 6-25

INDEX